Sanjit Debnath
Sisir Kumar Mitra
Sanjay Kumar Dutta Ray

Sustainable Litchi Cultivation

D1807554

Sanjit Debnath
Sisir Kumar Mitra
Sanjay Kumar Dutta Ray

Sustainable Litchi Cultivation

Current CO2 assimilation–tree water relations and nutrition programme optimization

LAP LAMBERT Academic Publishing

Impressum / Imprint
Bibliografische Information der Deutschen Nationalbibliothek: Die Deutsche Nationalbibliothek verzeichnet diese Publikation in der Deutschen Nationalbibliografie; detaillierte bibliografische Daten sind im Internet über http://dnb.d-nb.de abrufbar.
Alle in diesem Buch genannten Marken und Produktnamen unterliegen warenzeichen-, marken- oder patentrechtlichem Schutz bzw. sind Warenzeichen oder eingetragene Warenzeichen der jeweiligen Inhaber. Die Wiedergabe von Marken, Produktnamen, Gebrauchsnamen, Handelsnamen, Warenbezeichnungen u.s.w. in diesem Werk berechtigt auch ohne besondere Kennzeichnung nicht zu der Annahme, dass solche Namen im Sinne der Warenzeichen- und Markenschutzgesetzgebung als frei zu betrachten wären und daher von jedermann benutzt werden dürften.

Bibliographic information published by the Deutsche Nationalbibliothek: The Deutsche Nationalbibliothek lists this publication in the Deutsche Nationalbibliografie; detailed bibliographic data are available in the Internet at http://dnb.d-nb.de.
Any brand names and product names mentioned in this book are subject to trademark, brand or patent protection and are trademarks or registered trademarks of their respective holders. The use of brand names, product names, common names, trade names, product descriptions etc. even without a particular marking in this work is in no way to be construed to mean that such names may be regarded as unrestricted in respect of trademark and brand protection legislation and could thus be used by anyone.

Coverbild / Cover image: www.ingimage.com

Verlag / Publisher:
LAP LAMBERT Academic Publishing
ist ein Imprint der / is a trademark of
OmniScriptum GmbH & Co. KG
Heinrich-Böcking-Str. 6-8, 66121 Saarbrücken, Deutschland / Germany
Email: info@lap-publishing.com

Herstellung: siehe letzte Seite /
Printed at: see last page
ISBN: 978-3-659-71962-2

Zugl. / Approved by: Kolkata, Bidhan Chandra Krishi Viswavidyalaya (State Agriculture University), Diss., 2005

1.0 INTRODUCTION

Litchi is one of the most important subtropical fruits of the world. This fruit is very much popular due to its attractive consumer-appeal and its luring taste in sugar-acid blending and flavour.

The edible part (pulp) is called 'aril' which is very succulent, juicy and aromatic and account for 60 to 75% of fresh fruit weight at harvest. The moisture content in litchi fruit varies from 77 to 83% in different cultivars. The pulp contains 6.74 to 20.6 % sugars and 0.20 to 0.64% acidity, about 1.10% protein, 0.10% fat and considerable amount of calcium, phosphorus, vitamin C, B_1 and B_{12}. The ascorbic acid content varies from 40.2 to 80.8 mg/110g of pulp in different cultivars. Malic acid is the predominant nonvolatile acid in litchi (Maity and Mitra, 2001). The attractive peel colour of ripe fruit is due to the presence of anthocyanin which is comprised of cyanidin-3-rutinoside (>75%), cyanidin-3-glucoside (<17%) and malvinidin-3-acetylglucoside (<9%) pigments (Rivera *et al.*, 1999).

Litchi (*Litchi chinensis* Sonn.) belongs to the family Sapindaceae which includes some other fruit crops of horticultural importance i.e., longan (*Dimocarpus longan* Lour.), rambutan (*Nephelium lappeaceum* Linn.), pulsan (*Nephelium mutabile* Blume.) and jamaican akee (*Blighia sapida* Koening). (Groff, 1921, Maity and Mitra, 2001).

The litchi is indigenous to South-eastern China (Batten, 1984) and is considered to have reached eastern India through Myanmar (Burma) by the end of 17th century (Hayes, 1957). The leading litchi growing countries are China, India, Thailand, Madagascar, U.S.A., South Africa, Israel, Australia and Vietnam.

Although litchi has a long history in Asia, it is relatively new species in most countries and efforts to increase production have been relatively small compared with the more established tropical fruits such as citrus,

3

banana, mango and pineapple. The major factors limiting expansion of litchi production in India are narrow cultivar base, shortage of planting materials, poor growing techniques, lack of irrigation, high incidence of insect pests and poor post-harvest management (Ghosh, 2001). In general, the growing techniques are poor in Asia where most of the orchards are dependent on rainfall, while the Australian growers apply fertilizers based on soil and leaf tests, mostly irrigate the orchards, protect the orchard by netting and developed well established marketing groups. Presently, China is coming in a big way with thousands of hectare of closely planted new litchi orchards, heading towards more strong footing in the global litchi industry in near future (Mitra, 2004a). However, the expansion of litchi industry elsewhere in the world entitled research activity towards cultivar improvement, canopy management, water and nutrition management, control of pests and diseases and post-harvest technology and marketing.

India ranks second in the world after China in production of litchi. Among the fruit crops in India, the litchi has registered the highest increase in area (12.92%) and production (20.41%) during 1997-98 (Anon, 2001). The area and production of litchi in India during 2000-01 were estimated to be 53.6 thousand hectare and 412.0 thousand tonnes, respectively (Anon, 2003). The major litchi growing states in India are Bihar, Uttaranchal, West Bengal, Punjab, Assam and Tripura. However, three districts of Bihar i.e., Muzaffarpur, Saharanpur and Champaran account for 75 to 80 % of total production of litchi in India.

In West Bengal, litchi occupies about 6.36 thousand hectare with an average production of 63.92 thousand tonnes and 10.05 t/ha productivity (Anon, 2003-04). According to the information published by the Planning Board, Govt. of West Bengal, the growth of litchi during the last five years in the state was 59.40% in area and 70.54% in production (Anon, 2002-03). The litchi producing districts in West Bengal are Murshidabad, Malda,

Hoogly, North and South 24 Parganas and Nadia. However, further growth potential both in area expansion and increased productivity exists in the state. Besides, to accept the challenge for better exploitation of the export-potential of this fruit crop in the state.

The demand of litchi both in local market as well as in export market is increasing besides giving higher economic returns to the growers. Studies by the Agricultural and Processed Food Products Export Development Authority (APEDA) also revealed that Indian litchi is highly export competitive even when the competing countries are Australia, Mauritius, South Africa and Madagascar. Again, there is enough scope for the litchi growers of West Bengal to enter the markets of Punjab, Maharastra and Gujarat which are presently occupied by the litchi produced in Bihar.

Appropriate agricultural trade policies can take advantage of the global markets resulting in increased exports of agricultural products. These exports contribute to the increased foreign exchange earnings, faster economic growth and poverty alleviation in rural areas (Chandra Babu, 2002). The volume of export gradually increased during the last few years (5 tonnes, 2700 containers and 30 tonnes in 1993, 1994 and 1995, respectively) (Ray, 2001a). However, the estimated scope for export of Indian litchi is about 10,000 tonnes valued at Rs. 1,000 millions (Singhal, 1995), simply indicates the poor performance in export.

The harvesting and marketing season of major litchi producing countries in the Northern Hemisphere (India, China, Taiwan, Thailand, Vietnam, Florida, Israel, Pakistan, Indonesia, Philippines etc.) is usually between May and July. In the Southern Hemisphere (Australia, Madagascar, South Africa, Mauritius, Reunion, Brazil), it is between November and March with the peak period during December to January (Kanwar, 2002).

5

For fresh fruits, most of the consumer markets prefer large, highly coloured, sweet fruits with small seeds. However, it is essential to maintain the Codex Standard (Codex Alimentarius Commission, 1995) for exploring the International litchi markets (Ghosh, 2001). There is good potential to supply litchi in European, Gulf and Russian markets during May to July from India. Markets in Australia, New Zealand and South Africa can also be explored by Indian producers (Kanwar, 2002). At the National Seminar on "Advances in production and post-harvest technology of litchi for export", Mitra (2004a) visualized a large demand of fresh litchi fruit in the domestic markets, while production for export market is likely to face the stiff competition from the major global players like China, Thailand and Israel.

Yield from litchi orchards are often low and variable. The choicest commercial cultivar of litchi in West Bengal, Bombai, is also prone to this problem. The major associate factors of this problem are lack of suitable nutrition programme, especially the timing of nitrogenous fertilizer application for monitoring vegetative growth and dormancy, non-availability of a leaf nutrient standard for optimum yield and insufficient information on the effects of water deficits on plant water relation, CO_2 assimilation, growth and yield.

Research activity on litchi production technology generated valuable information befitting for the litchi growers. Fertilizer recommendations for bearing litchi trees based on yield record and tree size/age are available (Ghosh and Mitra, 1990, Mitra, 2004b). However, the soil and leaf analysis will provide suitable guidelines in perfecting the dose for targeted yield. Again, among the mineral nutrients, nitrogen plays the most important role in relation to vegetative growth and flowering in litchi and has a pronounced effect on the productivity of the orchard. Specific information are therefore, in demand not only to optimize fertilizer dose but also to standardize the time of fertilizer application in relation to vegetative flushing and leaf nutrient build up. Establishment of an adequate foliar nutrients range for optimum yield as well as refinement of leaf and soil sampling techniques are needed to develop a balanced fertilizer programme for bearing litchi orchards.

The significance of stress and non-stress periods of soil moisture levels on productivity of litchi orchards is also noteworthy. A dry period before flowering is beneficial to litchi flowering (Mitra, 2004b). But soil moisture deficits after panicle initiation caused reduced fruit set, poor fruit growth and more cracking of fruits (Oothuizen *et al.*, 1994, Menzel *et al.*, 1995a). Very little informations are available on the effect of water relation and CO_2 assimilation of litchi and how these impact on growth and yield of field grown litchi trees.

In view of the commercial importance of litchi in West Bengal and the results of investigations on litchi production technologies stated above, experiments were conducted for *"Optimizing nutrition programme and irrigation for sustainable litchi production"* in West Bengal with the following objectives.

I. **Timining of fertilizer N application** :To develop relationship between vegetative flushing and leaf N, timing of fertilizer application at different stages.

II. **Refinement of leaf and soil sampling** :To develop leaf and soil sampling techniques and leaf nutrient standards for sustainable production.

III. **Effect of water relations and CO_2 assimilation** : To study the effects of water relations, CO_2 assimilation on growth and yield of field grown trees.

7

2.0 REVIEW OF LITERATURE

2.1 Relationship between vegetative flushing and leaf N, timing of fertilizer N application

Litchi needs proper nutritional care for sustainable production. The yield of litchi varied from 0.62 t/ha in Thailand to 8.05 t/ha in India (Menzel and Simpson, 1986a). One of the reasons of such vast difference would be the difference in the nutritional care of plants (Rao *et al.*, 1985). The soil and foliar nutrients were reported to be more in high yielding than the low yielding plants. Fertilizer recommendation for bearing litchi trees based on yield record and tree size/age is available (Table 2.1.1), however, the soil and leaf analysis are expected to provide suitable guideline in perfecting the dose.

Table 2.1.1. Recommended N, P_2O_5 and K_2O rates for bearing litchi tree per year.

Country	Amount of nutrients (g/tree/year)			References
	N	P_2O_5	K_2O	
India	1,470	680	540	Nijjar (1981)
	600	250	750	Ray *et al.* (1985)
	600	200	600	Ghosh and Mitra (1990)
	600	200	600	Hasan and Chottapadhyay (1993, 1997)
Hawaii	763	327	633	Yee (1972)
South Africa	700	54	250	Langenegger (1974)
Australia	600	200	600	Menzel (1984)
China	800	640	320	Anon (1978)

Specific information are in demand not only to optimize fertilizer dose but also to standardize the time of fertilizer application in relation to vegetative flushing and leaf nutrient build up, because these have direct influence on the yield performance of a litchi tree. Among the mineral nutrients, nitrogen plays the most important role in relation to vegetative growth in litchi orchards and has a pronounced effect on the bearing of the trees. Although there was a strong correlation between yield and leaf N upto 14.7 mg N/g, very high rates of N depressed yield, presumably because of excessive vegetative flushing (Langenegger, 1975, Koen *et al.*, 1981, Koen and Smart, 1982).

Zhuang *et al.* (1988) showed that poor fruiting in litchi was associated with low levels of nitrate –N (< 960 μ g/g), P (< 0.5 mg/g) and K (< 11.6 mg/g) in the leaf prior to and after fruit set. Huang *et al.* (1998) developed a relationship between leaf N content and yield of cv. 'Haak Yip' in Taiwan and showed that fruit yield should be guaranteed to be > 90% of maximum as leaf N content remained in the range 1.5–1.9 % at anthesis.

Menzel and Simpson (1990a) observed that the concentration of leaf N fell slightly during the initiation of the vegetative flush, then increased to a maximum (1.8%N) when the flush matured. The value then fell to a minimum (1.4%N) during fruit development. Again, Menzel and Simpson (1992) opined that flushing did not appear to be affected by temperature and rainfall or by leaf N levels and the interactions of these factors demonstrated the difficulties in separating the effects of different environmental factors on vegetative growth in litchi. They also observed that the trees did not flush during winter or spring unless flowering or fruit set failed and there was no concurrent vegetative and reproductive growth in the same shoot. However, the understanding of possible influence of timing of fertilizer N application on leaf N content and growth flushing requires experimentation.

Seasonal variation in carbohydrate reserves had been studied by Menzel *et al.* (1995b). They found that about 50% of the total starch reserve were accumulated in the small and medium branches (representing 35% of the tree's biomass), compared with about 8% in leaves which accounted about 25% of tree's biomass. Starch concentrations were higher in all parts of a flowering tree than a vegetative tree. Maximum concentration of starch was recorded before anthesis followed by a decline during fruit development and harvest and remained low just after or during vegetative growth phase. The largest seasonal variation occurred in small branches, while it remained more stable in leaves, trunk and roots.

It is generally suggested that the mobile nutrients such as N, P and K should be applied well before fruit set to build up tree reserves for flowering and fruiting, while the less mobile elements will probably need to be supplied during flowering and fruiting (Menzel et al., 1992a).

However, till today, there seems to be some confusion on fertilizing (especially fertilizer N application) litchi trees around flowering over the World. In China, trees received a boost (either by side dressing or by foliar spray) just prior to flowering to strengthen the inflorescence and reduce fruit drop (Winks et al., 1983). In South Africa, half the amount of N and K is applied at flowering (Koen and Smart, 1983a). Paxton and Chapman (1980) recommended that trees in subtropical Australia must not be fertilized during spring, because it promoted growth flushes and premature flower and fruit abscission. In India, 3/4th of the total inorganic fertilizers are applied in June-July after harvest and rest 1/4th is applied in early April (Ray, 1991a).

The timing of fertilization, especially application of fertilizer N is of great concern because it has a direct effect on growth flushing (Menzel and Simpson, 1986b) and floral initiation (Winks et al., 1983). It was observed that the vegetative growth after September led to erratic bearing in litchi (Mustard and Lynch, 1959) and the vegetative flushing in the last week of November hardly produced any panicle in March. However, the new flushes may appear due to either late rain or high temperature, but they fail to bear flower in next spring.

Chen et al. (1994) also observed that panicle formation was highly correlated with the time of last flush in the previous year. The twigs which flushed in July or August had higher percentage panicle formation in the next year than those flushed in October. They also noted that if flush growth was >10cm, the leaves grew continuously through the winter and

10

no panicles developed in the following year. They suggested that the flush growth should be restricted to 0.6-2cm in October-November to prevent alternate fruit set and stabilize yield. These observations were also supported by the findings of Davenport *et al.* (1999). In contrary, Li *et al.* (1998) suggested to promote the last autumn shoots which sprout in early October to obtain good yield in litchi cv. 'Nuomizi'.

Nitrogen application in early spring can promote growth flushing which may compete with the flowers and increase fruit drop (Menzel and Simpson, 1986b), while application during autumn and winter may encourage excessive flushing prior to the normal period of flower initiation. Usually no fertilizer is applied during autumn and winter so as to restrict growth flushing and provide some sort of stress to the trees (Ray, 2001b). However, nutrient management for bearing trees is directed firstly to gain maximum vegetative growth immediately following harvesting in summer. The second requirement is to maintain the trees in a state of dormancy for 3–4 months before flowering. The third requirement is to maintain nutrition at a high rate once fruits have set, to prevent any stress (Menzel and Simpson, 1986b).

It has been reported that the time of fertilizer N application resulted in significant variation in some of the fruit characters (Das *et al.*, 2001) but had no effect on yield of litchi (Menzel *et al.*, 1994). Das *et al.* (2001) conducted an experiment with litchi cv. Bombai in West Bengal, on the timing of N fertilizer application with three levels of N (200, 400 & 600 g/plant/year and a fixed dose of P_2O_5 and K_2O at 400 and 600 g/plant/year, respectively), applied at four different times viz., entire dose in June (S1), ½ in June + ½ in August (S2), ½ in June + ¼ in August + ¼ in October (S3) and ¼ in June + ¼ in July + ¼ in August + ¼ in September (S4). The different time of N application showed no significant variation on fruit weight and yield/plant. However, the interaction effect of different levels of N and their time of application showed the maximum yield (70.74

kg/plant) by treatment with 600 g N/plant/year applied ½ in June + ½ in August.

2.2 Refinement of leaf and soil sampling.

Modern methods of interpretation of leaf analysis data such as Critical limit concept, Nutrient ratio, Crop logging, Diagnosis and Recommend-ations Integrated System (DRIS) and Boundary line concept, can diagnose the growth/yield limiting nutrients and provide guideline for recommendation for optimum use of nutrients (Bhargava, 1999). It was supposed that combined leaf and soil sampling (for 5 years) is needed to develop a balanced fertilizer programme for perennial fruit crops like litchi.

However, the mineral contents of leaf had been reported to vary due to variation in age of leaf (Menzel *et al.*, 1987, Sanyal and Mitra, 1990), position of shoots (Sanyal and Mitra, 1990) and time/stage of sampling (Menzel *et al.*, 1988, Nijjar, 1981). In general, the levels of N, P, K, Zn and Fe were lower in fruiting compared to non-fruiting branches while the opposite was true for Ca, Mg, Mn and B (Menzel *et al.*, 1992a).

It has also been showed that the main effect of fruiting was to reduce leaf K and to a smaller degree leaf N, P, Zn and Na and to increase Ca, Mg, Mn and B (Menzel *et al.*, 1988). While the response of N and Cu depended on the level of supply. In case of older leaves the levels of P, K and Zn declined while those of Ca, Mg, Na, Cl, Fe, Mn and B increased. However, Sanyal and Mitra (1990) observed that the concentration of N, P and K increased and that of Ca, Mg and S decreased with leaf age.

2.2.1 Sampling procedure

Index tissue: Leaf is the principal site of the crop physiological activity, showing symptoms of nutrient stress and hence is an ideal index tissue. It has also been reported that the concentration of nutrients at specific stages of growth are related to the tree performance (Lagatu and Maume, 1926). Precisely, for these reasons emphasis is now placed to correlate the foliar nutritional status of litchi with yield. To make full use of this tool, answer to some basic questions like leaf age, position on the tree / direction and sample size etc. should be known so that a period, when most of elements in the leaf were stable, could be found for sampling. The sampling error must be essentially of the least possible magnitude (Koen and Smart, 1983b).

There is considerable variation with regard to the period of most nutrients stability in different litchi growing areas as also with different varieties (Kotur and Singh, 1992a), probably with the sampling procedure also (Table 2.2.1).

Table 2.2.1. Leaf sampling procedure of litchi followed in different countries.

Country	Sampling techniques	References
Australia	Mature leaf behind the fruiting cluster sampled in October, prior to fertilization in November (i.e., 1–2 weeks after panicle emergence)	Menzel *et al.* (1992b)
Florida	Leaves are taken from non-flowering and non-flushing terminals during mid-flowering (March)	Young and Koo (1964)
South Africa	The second leaf (6-7 month old) below the developing fruit cluster (6-8 weeks after fruit set) is sampled, taking the two central leaflets with the midribs and stalks	Koen and Smart (1983b) Menzel and Simpson(1987)
India	i) In Punjab, leaves were sampled prior to floral initiation	Nijjar (1981)
	ii) The 1st and 2nd pairs of leaves (for N, P & K) and 5th and 6th pairs of leaves (for Ca, Mg & S) from vertical shoots from either East, West, North or South parts of the crown sampled in August or September were suitable for estimating nutritional status of cv. Bombai trees	Sanyal and Mitra (1990)
		Kotur and Singh

growing in alluvial plains of W.B.	(1992b)
iii) 6-7 months old leaves from all directions from mid tree height from Sept. flush which coincided with peak flowering period	Bhargava (1999)
iv) Second pair of leaflets from tip from autumn flush (6 month old) with sample size of 40	

Regarding the size of leaf sample, Azad (1972) recommended 25-65 middle leaflets from fruiting terminals while Chaudhury and Singh (1993) suggested 25–40 leaflets collected from 4–7 months old leaves between November to February before flowering. Bhargava (1999) considered that leaf sample size of 40 is ideal.

Interpretation of leaf analysis is not a simple job. The laboratory values have to be compared with standard values so that nutrients in short/excess supply could be identified. The optimum foliar levels of any given nutrient element tend to vary with climate, cultural practices and cultivars and these must be considered carefully while drawing any conclusions. Decision making is complicated further by the fact that there is no strict proportional relationship between the amount of particular nutrient element contained in the leaf and the magnitude of corrective measures to be taken. Therefore, it is not possible to establish general standards to help in decision-making process. Probably for these reasons, the adequate foliar nutrient levels vary in different countries (Table 2.2.2).

The Diagnosis and Recommendations Integrated System (DRIS) has been used as an alternative approach for foliar diagnosis of micronutrients in litchi by Hundal and Arora (1996). In their opinion, DRIS norms were better than the diagnosis made by the sufficiency range method. DRIS reflected nutrient balance and identified the order in which nutrients were likely to be limiting. Results of these workers showed that the DRIS norms could be used irrespective of cultivar and position of leaf sampled from the floral or non-flower panicles. According to the DRIS approach, the inadequacy levels for Zn, Cu, Fe and Mn in litchi were 14, 10, 190 and 20 ppm compared with 15-30, <10, <50 and <100 ppm, respectively by the sufficiency range method as suggested by Menzel *et al.* (1992c).

Table 2.2.2.Adequate nutrient range in the leaves of bearing litchi tree.

Element	Israel	Australia	South Africa	India
N	1.5-1.7%	1.5-1.8%	1.47-1.52%	1.70-1.80%
P	0.15-0.30%	0.14-0.22%	0.15-0.18%	0.20-0.25%
K	0.70-0.80%	0.70-1.10%	0.90-1.05%	1.00-1.10%
Ca	2.00-2.30%	0.60-1.00%	-	0.80-1.00%
Mg	0.35-0.45%	0.30-0.50%	-	0.40-0.50%
S	-	-	-	0.20-0.25%
Na	0.30-0.50%	<500 ppm	-	< 300 ppm
Cl	-	<0.25%	-	-
Fe	40-70 ppm	50-100 ppm	-	80-100 ppm
Mn	40-80 ppm	100-250ppm	-	180-200ppm
Zn	12-16 ppm	15-30 ppm	-	15-25 ppm
B	45-75 ppm	25-60 ppm	-	25-40 ppm
Cu	-	10-25 ppm	-	10-15 ppm
References	Kadman and Slor (1982)	Menzel *et al.* (1992c)	Koen *et al.* (1981)	Mitra (2004b)

2.3 Effect of water relations and CO_2 assimilation on growth and yield of field grown litchi tree.

For successful cultivation of litchi, assured and adequate moisture supply is essential. With a well distributed rainfall between 1250-1550 mm, litchi can be grown even without irrigation. Such areas do exist in China (Groff, 1943). However, in many litchi-growing countries, such as Israel, South Africa and Australia, rainfall and humidity during flowering and fruiting are very low, and unirrigated trees should be expected to develop significant water deficits. In north-India where rainfall is restricted mainly between June and September, litchi is grown successfully with supplemental irrigation during fruiting period.

2.3.1 Stress and non-stress periods

Litchi trees require adequate soil moisture during fruit setting and subsequent development of fruits till maturity. The high soil moisture level during maturity reduces skin cracking (Sharma and Ray, 1987). Any stress period during flowering to maturity may cause detrimental effects such as poor fruit set (Hasan and Chattapadhay, 1991), reduced photosynthesis, leaf wilting and dropping, poor flowering, small and cracked fruits, excessive fruit drop and even tree death (Oosthuizen et al., 1994, Roe and Oosthuizen, 1994). Menzel et al. (1995a) also reported that in 10 years old 'Tai So' litchi water deficits reduced initial fruit set by 30% and final fruit set by 70% and increased fruit splitting by 30-40 %. Water deficits did not alter the sigmoidal pattern of growth but reduced yield from 51.4 ± 5.5 kg/tree in well-watered trees to 7.4 ± 3.3 kg/tree in droughted trees.

However, high soil moisture prior to floral initiation promoted vegetative growth and suppressed flowering, while the low soil moisture status during that period restricted vegetative growth and promoted flowering in Israel (Stern et al., 1990, 1993). A pre-flowering dry period (with-holding irrigation in autumn and winter) appeared beneficial to litchi flowering under the climatic condition of West Bengal (Mitra, 2004b).

2.3.2 Root distribution and pattern of moisture removal

For effective management of root zone soil moisture, it is essential to establish the rooting depth of the specific crop as it greatly varies with crop species and age. Roots of fruit trees, in general, extend down to 1.5 m (Bielorai et al., 1973). However, litchi trees were found to have a shallow root distribution when grown in subtropical Queensland. Although some roots were found down to a depth of 100 cm, most roots were located in the top 20-60 cm of the soil. Some roots were found at greater depths but their contribution to the feeder root system was < 3%. The highest feeder root density lies in the 0-30 cm soil layer. The depth of rooting was greater in fine textured soil compared with coarse textured soils but the total amount of roots in the profile showed the reverse trend (Menzel *et al.*, 1990).

When the water potential throughout the root zone is near field capacity, roots will absorb water rapidly from the upper parts of the soil where O_2 is most abundant. As the soil dries, the soil water potential in the surface soil decreases, water uptake will shift to deeper soil layers, where the supply is less but soil is moist and the potential is high. Thus the root zone is progressively depleted of available soil moisture (ASM). After rewetting, water absorption again shifts back towards surface soil layers.

2.3.3 Determining irrigation requirements

In general, there are different approaches for determination of irrigation requirements in litchi. The useful approaches are as follow :

i) Measurement of soil water status : Soil water content is measured in several ways viz. Gravimetric analysis, by using Tensiometer or a Resistance block or Neutron moisture probe.

ii) Using soil moisture data for irrigation : Here the main aim is to maintain the moisture reserve of the root zone to a predetermined level to serve as the minimum allowable level (i.e., the "time to refill" criterion). Soil moisture determination also helps in estimating crop evapotranspiration (ET) through the water balance techniques :

$ET = (P+I+U) – (R+D) \pm \Delta$ ASM, where P = precipitation, I = irrigation, U = upward capillary flow into root zone from below, R = run-off, D = down ward drainage and $\pm \Delta$ ASM= change in soil moisture over a given time of interval.

As a rule of thumb, water is applied when 50% of the available water in the root zone depleted (Hillel, 1987). Hasan and Chattopadhyay (1990, 1992) also observed that yield decreased as the ASM depletion level increased by irrigation interval. Though the highest water use efficiency was observed with irrigation at 45% ASM depletion, the best yield could be harvested with only 30% ASM depletion.

iii) Plant based measurement : In addition or as an alternative to monitor the soil moisture, it is possible to monitor the water status of the plants. This can be done visually, as well as instrumentally, to detect early sign of thirst (incipient stress) in time to irrigate and thus prevent any significant reduction in yield.

In recent years measurement of leaf water potential (LWP) has been widely used for evaluating water status of the plants. However, the dynamic nature of LWP may pose great difficulties in reaching at any definite conclusion due to – a) very high diurnal variation in LWP (-0.1 MPa to –1.0 MPa) (Menzel and Simpson,1986c), b) changes in leaf–air water conductance by high degree of exposure of leaves to direct radiation (Menzel and Simpson, 1986c), c) osmotic adjustment in response to varying water status in the soil and air (Syvertsen, 1985, Batten *et al.*, 1992) and d) age/position of the sampled leaves. Usually, the pre-dawn values (-0.14 to –0.37 Mpa) of LWP, which are almost stable and free from influences of weather, are preferred (Menzel and Simpson, 1986c).

2.3.4 SPAC system of irrigation management

In the last two decades, the new concept of irrigation management i.e., *"Soil-Plant-Atmosphere-Continuum"*(SPAC) system has been developed to deduce a more precise and better controlled irrigation schedule for a particular crop, attaining maximum water use efficiency.

According to the SPAC system, the movement of water takes place from higher (soil) to lower (air) potential gradient (Table 2.3.1).

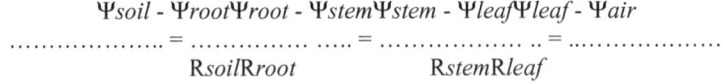

(Water molecule moves from soil to air, where the order of potential gradient is : soil > root > stem > leaf > air).

Table 2.3.1.Magnitude of water potential in soil-plant-atmosphere-continuum system (SPAC)*.

Component	Water potentials (bars)
Soil	-0.1 to –20.0
Leaf / plant tissues	-5.0 to-50.0
Atmosphere	-100 to –2,000

*Source : Plaut and Moreshet (1973)

The advent of new irrigation system (sprinkler, dripper or micro-sprayer etc.) has made it possible to establish and maintain soil moisture condition at more nearly optimum level.

Methods of irrigation: The methods of irrigation may be classified broadly as follows :

1) Surface method (Basin, Border strip, Furrow)
2) Sub-surface
3) Micro-irrigation (Dripper, sprinkler)

Each system has advantages and disadvantages and the selection of a method depends on the physical as well as economic factors associated with. Drip irrigation is an advanced method of irrigation which is preferred to other conventional methods (Table 2.3.2). However, care should be taken for the following limitations in drip irrigation system operation may encounter –

i) Emitter clogging due to biological factors (slimes/algae etc.), physical factors (sand, clay, plant fragments etc.) or chemical precipitation ($CaCO_3$) and reactions.

ii) Salinity built up in root zone.

iii) Physical or biological (rodents/rabbits/dogs etc.) damage to trickle lines.

iv) Economic and technical limitations, if any.

The components of drip irrigation system are :

1) Water treatment (supply pump, storage tank, fertilizer mixing unit, primary filter, pressure regulator, purification etc.)

2) Water transport (distribution pipe line, solenoid & flush valves, back-flow prevention)

3) Water emission (point source emitter, line source emitter or combination)

Bredell *et al.* (1975) used microjet sprinkler for application of herbicide in litchi orchards to keep the tree trunk out of reach of the spray. Menzel and Simpson (1986c) suggested that the over head type sprinkler is suitable for improving productivity of litchi trees where atmospheric humidity is low and wind speed is usually normal, as it decreased leaf-air water vapour concentration gradient. Drip irrigation has been employed successfully in Israel and South Africa.

Table 2.3.2.A comparative study of surface and drip irrigation system*.

Parameters	Surface irrigation system	Drip irrigation system
1. Water requirement	More (eg., for Papaya – 228 cm, Grape – 53 cm, Kinnow – 22.10 cm)	Less (eg., for Papaya – 73cm, Grape – 28 cm, Kinnow – 17.30 cm)
2. Pressure requirement	Ten times more(to irrigate 4 ha, 10 HP pump set is required)	One-tenth (to irrigate 4 ha, 1 HP pump set is required)
3. Evaporation loss	15 – 20 %	Almost nil
4. Labour requirement	More	Less
5. Weed growth	More	Less
6. Maintenance of uniform moisture	Practically impossible	Soil moisture is maintained almost

at root zone		at field capacity
7. Land leveling	Is a must in flood irrigation	Not essential
8. Application of agro-chemicals	Always not possible and economic	Easy and economic
9. Possibility of using salt water	May cause detrimental effect	Can be used up to a certain extent
10. Annual electricity charge	More	Less
11. Irrigation efficiency	35 – 50 %	90 – 95 %

*Patel (1999)

In India, micro-irrigation system was introduced at farmers' level around 1980, since then it has picked up momentum from 1,500 ha in 1985 to 0.3 million ha under micro-irrigation in 2000 (Singh *et al.*, 2000). Research activities in this field were undertaken by ICAR, SAUs through AICRP, DRIPNET projects or AP cess adhoc schemes at different agroclimatic conditions (Table 2.3.3).

Although there is very little information available on the performance of drip-irrigation in litchi, it has been found very much encouraging for other fruit crops like banana, grapes, papaya, citrus, guava, pomegranate, ber etc. (Singh *et al.*, 2000). Singh *et al.* (2000) reported higher yield and better water use efficiency (WUE) with drip irrigation in different fruit crops over surface method of irrigation. Drip irrigation caused highest WUE of 34.80 q/ha/cm in papaya at Kalyani followed by grape (11.60 q/ha/cm) at NCPA.

Shirgure *et al.* (2001) evaluated the effect of micro-irrigation system in comparison with the basin method of irrigation in Nagpur mandarin and recorded that micro-irrigation system resulted in better tree growth, higher yield and superior fruit quality. Microject 300^0 (2/plant) caused highest fruit yield (28.0 kg/tree), weight of individual fruit (162.5 g), TSS (7.83^0B) and juice content (47.82%) compared with 22.0 kg fruit/tree in microject 180^0 (2/plant) and 21.8 kg fruit/tree in basin irrigation system.

Table 2.3.3. Centres and crops covered under DRIPNET*

Institute	Crops
CISH, Lucknow	Mango, guava
NRC for Citrus, Nagpur	Citrus
NRC for Banana, Trichy	Banana
Mahatma Phule Krishi Vishwa Vidyalaya, Rahuri	Pomegranate and ber
NRC for Grapes, Pune	Grapes
NRC for Arid Horticulture, Bikaner	Arid fruits
NRC for Oil Palm, West Godavari	Oil palm

*Singh *et al.* (2000)

2.3.6 Tree water status and CO_2 assimilation in litchi

There is very little information on the effect of water relation and CO_2 assimilation of litchi and how these impact on growth and yield of field grown litchi trees. In a greenhouse experiment with potted plants of cv. 'Wai Chee', net CO_2 assimilation (A) declined as midday leaf water potential (Ψl) fell below $- 1.0$ Mpa and approached zero at a Ψl of about -4.0 Mpa when plants wilted (Chaikiattiyos *et al.*, 1994). In another experiment with 10 years old 'Tai So' litchi trees grown on a sandy loam soil and were watered weekly (well-watered treatment) or droughted from late July until January (draught treatment), Menzel *et al.* (1995a) recorded that at week 9, both stomatal conductance and net CO_2 assimilation (A) declined due to drought treatment (Table 2.3.4).

Roe *et al.* (1995) investigated the effect of water deficits on CO_2 assimilation in young 'Tai So' litchi trees growing in pots with sand or clay soil. In general, there was a decline in leaf water potential (Ψl), stomatal conductance (gs) and net CO_2 assimilation (A) measured at 0900 h as plants went without water for 0, 5, 15, 19 or 22 days. Due to water deficit, the net CO_2 assimilation rate fell to 18% of the value in well-watered plants, when Ψl declined to -3.2 Mpa. Variation also existed in Ψl and A due to soil type (Table 2.3.5).

Roe et al. (1997) also investigated the effect of current CO_2 assimilation rate and stored reserves on fruit growth of 12 years old cv. Mauritius litchi trees. They observed that fruit set was virtually eliminated in branches without leaves and therefore, they claimed that stored reserve alone cannot support a crop, rather a successful fruit set in litchi depends on the current CO_2 assimilation. It was also noted that the number of fruits/panicle and relative fruit retention (%) showed a significantly inverse relationship with the number of leaves removed. Retention of 30 leaves per panicle caused highest fruit retention of 27.2%, while fruit retention was as low as 2.7% only when all the leaves were removed from the bearing branch.

Table 2.3.4. Effect of tree water status on stomatal conductance and net CO_2 assimilation in litchi*

Treatment	Stomatal conductance (mmol CO_2 m^{-2}s^{-1})	Net CO_2 assimilation (μmol CO_2 m^{-2}s^{-1})
Well-watered treatment	70 - 300	3 – 13
Drought treatment	50 - 180	2 - 6

* (after Menzel et al., 1995a)

Table 2.3.5. Variation in leaf water potential (Ψl) and net CO_2 assimilation (A) due to plant water status and soil type*.

Treatment	Leaf water potential (Ψl) (Mpa)	Net CO_2 assimilation (μmol CO_2 m^{-2}s^{-1})
Well-watered plants	- 0.8	3 – 8 (clay) 4 – 12 (sandy)
Wilted plants		
Clay soil (after 17 day without water)	- 4.5	Almost nil
Sandy soil (after 4 days without water)	- 3.5	Almost nil

*(after Roe et al., 1995)

3.0 MATERIALS AND METHODS

The experiments were conducted at the Horticultural Research Station of Bidhan Chandra Krishi Viswavidyalaya, Mondouri, Nadia, West Bengal, India during 2001 to 2005 and funded under National Agricultural Technology Project from India Council of Agricultural Research.

3.1 Location of the experimental site

The Research station is situated at 9.75 m above sea level and the latitude and longitude are 23.5 ^{0} North and 89 ^{0} East, respectively.

3.2 Agro-climatic condition

The National Bureau of Soil Survey and Land Use Planning delineated this region as agro-ecological region of "Assam and Bengal Plains, hot humid". The place has a subtropical climate and the long term average (LTA) of annual rainfall was 159.48 cm and mean temperature ranged from 17.65^0 C to 30.78^0 C. There are three main seasons in this region namely summer (March to June), rainy (July to October) and winter (November to February). The long term average (LTA) of meteorological data(monthly mean of 1957 to 1991) is presented in Table 3.2.1 and details of the climatic conditions (monthly and weekly average) during the period of investigations are presented in Tables 3.2.2 and 3.2.3 (a,b). The source of meteorological data is the AICRP project on Agril. Meteorology, BCKV.

Table 3.2.1. Long term average (LTA) of meteorological data .*

Month	Temperature (^{0} C)			Relative humidity (Average) (%)	Total Rainfall (mm)	Total Evaporation (mm)
	Max	Min	Average			
January	25.33	9.97	17.65	64.28	18.73	97.82
February	27.88	13.33	20.60	60.61	14.66	136.59
March	32.95	19.89	26.42	57.42	39.01	227.51
April	36.81	23.89	30.35	59.12	52.63	268.74
May	36.43	25.13	30.78	66.67	107.47	277.85
June	36.40	25.12	30.76	78.47	361.47	217.35
July	32.34	26.68	29.51	83.29	309.96	162.80
August	32.66	24.80	28.73	83.05	272.44	135.24
September	32.74	25.51	29.12	82.72	270.25	115.52
October	30.28	23.77	27.02	76.75	126.82	112.39
November	29.95	16.27	23.11	67.85	19.39	94.94
December	25.71	11.34	18.52	63.87	1.98	86.93

24

Table 3.2.2. Monthly average of temperature, relative humidity and rainfall data during the period of investigations.

Month	Temperature (°C)			Relative humidity (%)			Rainfall (cm)
	Max	Min	Mean	Max	Min	Mean	
2001							
September	33.07	25.29	29.18	98.17	78.83	88.50	33.32
October	32.40	23.64	28.02	99.26	73.29	86.27	20.18
November	30.50	19.87	25.18	98.77	62.07	80.42	0.35
December	27.00	11.88	19.44	99.61	51.32	75.46	0.00
2002							
January	26.20	12.60	19.40	98.80	55.30	77.05	2.08
February	28.90	14.30	21.60	98.40	43.00	70.70	0.00
March	32.74	18.70	25.72	95.50	43.75	69.62	0.50
April	34.55	23.07	28.81	92.73	59.87	76.30	14.50
May	34.34	25.09	29.71	92.06	59.87	75.96	10.70
June	34.11	25.45	29.78	96.43	74.43	85.43	36.28
July	34.06	26.40	30.23	95.39	76.58	85.98	25.44
August	32.50	25.50	29.00	97.70	81.90	89.80	32.02
September	32.90	25.09	28.99	97.80	76.40	87.10	16.49
October	32.40	22.12	27.26	96.81	67.71	82.26	10.26
November	29.89	17.60	23.74	98.50	61.33	79.91	7.74
December	27.41	12.44	19.92	99.77	54.56	77.21	0.00
2003							
January	23.7	9.5	16.60	100	58	79.0	0.00
February	29.2	15.9	22.55	98	53	75.5	0.00
March	32.0	19.1	25.55	95	51	73.0	6.42
April	36.2	25.0	30.60	92	53	72.5	0.60
May	36.5	24.9	30.70	93	58	75.5	8.14
June	34.8	24.9	29.85	94	74	84.0	36.17
July	33.4	25.0	29.20	97	76	86.5	29.52
August	33.4	25.0	29.20	98	78	88.0	15.57
September	33.4	25.5	29.45	99	80	89.5	16.20
October	31.9	23.9	27.90	99	79	89.0	17.28
November	29.6	16.7	23.15	100	55	77.0	0.2
December	26.2	13.2	19.70	99	57	78.0	2.02
2004							
January	24.2	11.5	17.85	99	57	78.00	0.16
February	29.6	14.3	21.95	97	40	68.50	0.00
March	35.3	20.7	28.10	94	43	68.50	1.08
April	36.2	24.0	30.10	94	59	76.50	11.20
May	37.4	25.9	31.65	89	55	72.00	10.46

Table 3.2.3(a) : Weekly average of temperature, relative humidity during the period of fruit growth and development.

Standard meteoro-logical weeks		Month	Date	Temperature (°C)			Relative Humidity (%)		
				Max	Min	Average	Max	Min	Average
2003	10	March	5-11	30.1	13.9	22.0	90	37	63.5
	11	March	12-18	29.9	20.3	25.1	96	70	83.0
	12	March	19-25	33.1	19.9	26.5	97	48	72.5
	13	Mar-Apr	26-1	33.7	21.4	27.6	96	55	75.5
	14	April	2-8	35.7	24.6	30.2	94	53	73.5
	15	April	9-15	31.3	24.9	28.1	89	47	68.0
	16	April	16-22	36.3	25.2	30.8	93	59	76.0
	17	April	23-29	36.9	24.5	30.7	91	54	72.5
	18	Apr-May	30-6	36.1	24.1	30.1	93	57	75.0
	19	May	7-13	35.5	23.9	29.7	95	62	78.5
	20	May	14-20	36.7	24.7	30.7	94	54	74.0
	21	May	21-27	36.3	24.6	30.5	90	62	76.0
2004	10	March	5-11	32.1	16.8	24.5	92	43	67.5
	11	March	12-18	35.5	20.9	28.2	97	45	71.0
	12	March	19-25	37.7	24.1	30.9	97	46	71.5
	13	Mar-Apr	26-1	36.6	24.5	30.6	90	48	69.0
	14	April	2-8	35.5	21.9	28.7	93	57	75.0
	15	April	9-15	36.6	25.6	31.1	93	55	74.0
	16	April	16-22	37.8	25.2	31.5	93	57	75.0
	17	April	23-29	35.2	23.5	29.3	96	64	80.0
	18	Apr-May	30-6	36.8	25.2	31.0	87	45	66.0
	19	May	7-13	38.8	28.1	33.4	84	45	64.0
	20	May	14-20	39.0	26.5	32.7	91	55	73.0
	21	May	21-27	35.0	23.5	29.2	92	69	80.0

Table 3.2.3(b) : Weekly average of rainfall, evaporation and vapour pressure during the period of fruit growth and development.

Standard meteoro-logical weeks		Month	Date	Total Rainfall (mm)	Total Evapor-ation (mm)	Vapour Pressure (kPa)	
						At 7LMT	At 14 LMT
2003	10	March	5-11	0.0	31.3	1.516	1.615
	11	March	12-18	30.1	23.3	2.369	2.454
	12	March	19-25	10.4	22.5	2.447	2.567
	13	Mar-Apr	26-1	23.7	31.6	2.745	2.682
	14	April	2-8	0.0	29.5	3.128	3.012
	15	April	9-15	0.0	32.5	3.059	2.835
	16	April	16-22	3.2	29.9	3.79	3.453
	17	April	23-29	2.2	35.1	3.129	3.100
	18	Apr-May	30-6	21.1	31.1	3.155	3.078
	19	May	7-13	11.8	25.6	3.177	3.186
	20	May	14-20	0.0	33.4	3.306	3.176
	21	May	21-27	48.5	34.0	3.202	3.469
2004	10	March	5-11	10.4	26.6	1.882	1.762
	11	March	12-18	0.0	23.6	2.586	2.376
	12	March	19-25	0.0	29.5	3.048	2.778
	13	Mar-Apr	26-1	0.4	26.1	2.921	2.658
	14	April	2-8	53.7	31.5	2.837	2.930
	15	April	9-15	0.0	37.6	3.245	3.211
	16	April	16-22	19.1	34.3	3.256	3.401
	17	April	23-29	38.0	31.2	2.954	3.122
	18	Apr-May	30-6	1.2	30.9	3.053	2.623
	19	May	7-13	0.0	43.7	3.400	2.836
	20	May	14-20	47.1	38.2	3.591	3.495
	21	May	21-27	38.5	44.7	2.940	3.377

Table 3.2.4. The physico-chemical properties of soil of the experimental site.

Physical properties					
Mechanical composition	:	Sand-55.40%, Silt-23.00%, Clay- 21.60%			
		50 cm	100 cm	150 cm	Average
Field capacity (% v/v)	:	26.10	23.40	29.60	26.37
Bulk density (g/cc)	:	1.44	1.50	1.51	1.48
Chemical properties					
pH : 6.80			Organic carbon : 0.65		
Total nitrogen : 0.06%			Available phosphorus : 29 ppm		
Available potassium : 42 ppm					

3.3 Experimental trees

The experiments I and II were carried out on 22 years old litchi trees of cv. 'Bombai'. The orchard of 'Bombai' litchi was planted in the year 1978 following square system of planting at a spacing of 7.5 m X 7.5 m. The experiment III was conducted in a 15 years old orchard of 'Bombai' litchi planted at a spacing of 7.5 m X 7.5 m of the same research station.

3.4 Orchard management practices

The orchard was ploughed twice a year, before and after the monsoon. The space around the trees was kept clean by occasional weeding and spading. The fallen leaves served as natural mulch. The dose of fertilizer was guided by the initial test values of soil and tree nutritional status. The dose applied was N, P_2O_5 and K_2O @ 400, 200 and 400 g/tree/year, respectively applied in two splits i.e., after fruit harvest (July) and after fruit set (March), except the trees under fertilizer trial.

The plant nutrients were supplied from urea, single super phosphate and muriate of potash. Fertilization was followed by irrigation. During fruit growth period two to three irrigation with 5-6 cm depth of irrigation were applied. However, in the irrigation experiment, irrigation was applied following the technical programme.

Spray schedule followed to protect the newly emerged vegetative growth, flowering panicle and developing fruits from insect pest attack are as follows :

Post harvest vegetative growth period(June – October)	: Sprayed two to three times depending upon the intensity of leaf roller infestation with Carbaryl 50% WDP @ 2.5 g/l, Acefate 75% WP @ 1g/l, Dichlorovos 76% EC @ 0.75 ml/l
Flowering and fruiting period (February - May)	: Sprayed three to five times depending upon the intensity of insect pest infestation with Endosulfan 35% EC @ 2 ml/l + Carbendazin 50% WP @ 1g/l, Acefate 75% WP @ 1g/l, Dichlorovos 76% EC @ 0.75 ml/l, Carbaryl 50% WDP @ 2.5 g/L and Malathion 50% EC @ 2 ml/l.
Dormancy period (November–January)	: Pruned the infested branches and sprayed with Dichlorovos 76% EC @ 0.75 ml/l or Endosulfan 35% EC @ 2 ml/l to control bark eating caterpillar.

3.5 Experimental details

Experiment I : Timining of fertilizer N application

Objective : To develop relationship between vegetative flushing and leaf N, timing of fertilizer N application at different stages.

Methodology : Nitrogen was applied at 400 g/tree/year while a fixed rate of P_2O_5 at 200 g/plant/year and K_2O at 400 g/plant/year were applied after harvest (July) and after fruit set (March) in two equal splits.

Time of application :

i.   After fruit harvest (July) (T_1)

ii. In two splits, after harvest (July) and after fruit set (March) (T_2)

iii. In two splits, after harvest (July) and in autumn (September) (T_3)

iv. In three splits, June, September and March (T_4)

v.   Conventional method (as control) in two splits, onset and the end of monsoon. (T_5)

Replication : 4

Unit trees/replication : 4

Design : Randomized blocks

Tree age : 22 years, cv. Bombai.

Observations :

a. Monthly proportion of the terminal shoots with vegetative flushing and flowering.

b. Details of flowering, yield and fruit quality.

c. Monthly estimation of N content of leaf and annually after fruit set for P, K, Ca, Mg, Fe, Zn, Cu and B.

d. Carbohydrate reserves in leaf and shoots at pre-flowering stage (December) and after fruit set (March).

e. The relation between pre-flowering (December) plant nutrient status and flowering.

f. The relation between plant nutrient status at flowering (March) and yield and quality of fruit.

Experiment II : Refinement of leaf and soil sampling

Objectives :

i. Study of leaf variables to standardize the sampling of leaves : a) direction (east, west, north, south of the tree), b) leaf age/position of leaf on the shoot (first, second, third, fourth, fifth and sixth leaf from the shoot tip, taking the second pair of leaflets from each leaf) and c) canopy height (1m, 2m, 3m, 4m, 5m and >5m from the ground level).

ii. Development of leaf nutrient standards for optimum yield and recommendation for sustainable production.

iii. The levels of soil nutrient contents at different depth at different growth and development periods.

Methodology : Based on yield performance of the trees of experiment-I, the trees under treatment 2 (highest yield in 2001) were selected for experiment-II.

Replication : 4

Unit trees/replication : 4

Design : Randomised blocks

Experiment-III : Effects of water relations and CO_2 assimilation on growth and yield of litchi tree.

Objective : To study the effects of water relations and CO_2 assimilation on growth and yield of field grown litchi trees.

Methodology : Basic information were recorded onBulk density and average soil water content (θ) form 0-150 cm at field capacity.

Trees were irrigated with two drippers per tree delivering 8 l h^{-1} dripper^{-1}. Soil water content was measured weekly with a neutron moisture probe (503 Hydroprobe, CPN, USA). Water use (E_t) in the experimental plots was calculated from the change in average θ from 0-150 cm each week, assuming no deep drainage below 150 cm.

Leaf water potential (Ψ_L) of a single leaf per tree was measured on the shaded side of each tree at 0900 h and on the sunlit side of each tree at 1400 h with a pressure chamber. Samples were collected weekly (only from week 5 for the afternoon sampling). Stomatal conductance (gs) and net CO_2 assimilation rate (A) were measured with a photosynthesis meter (CI 310, CID, Inc.).

Tree age : 15 years of cv. Bombai.

Total number of trees employed for the experiment : 60

Replication : 4

Unit trees/replication : 3

Design : Randomised blocks

In this experiment, soil moisture was supplemented from anthesis through drip irrigation system at 80, 60, 40, 20 and 0% of pan coefficient (i.e., five treatments). The dynamics of soil moisture content and tree-water status were studied at weekly interval. The variations in shoot growth, net CO_2 assimilation rate, fruit set, retention and development of fruit and ultimate yield and quality of fruit at different irrigation levels were recorded. The effects of water stress at different stages of fruit development and its possible role on fruit cracking (a major problem of litchi growers in the state) were investigated.

3.6 Details of methodology for observations

3.6.1 Tree growth study : Before imposing treatments, initial tree height (m) and crown diameter (m) (North-South & East-West) were measured and tree volume (m^3) was calculated with the use of formula $4/3\Pi\ a^2\ b$, where a=half of spread, b=half of height.

Monthly increase in shoot length of tagged shoots was recorded between $15-17^{th}$ of each month and the percentage increase over the previous month was calculated.

3.6.2 Study of flowering details : Representative panicle samples were studied for panicle dimension and sex ratio. Initial fruit set, dropping and splitting of fruits in the following weeks were recorded from tagged panicles.

3.6.3 Study of weekly variation in peel, seed, aril and fresh weight of fruit : Fruit samples were collected weekly from 4th week after fruit set till harvest for recording peel, seed, aril and fresh weight of fruit. Weekly variation in aril percentage for respective treatments was calculated.

3.6.4 Study of fruit yield : Average number of fruits per panicle, number of panicles per square meter of canopy area were recorded and multiplied by the average fruit weight for recording fruit yield per tree.

3.6.5 Study of physico-chemical parameters of fruit : Twenty representative samples from each replication were used for measuring fruit size, weight of fruit, stone and aril. The same samples were used for chemical analysis of fruit quality parameters. The juice was extracted from fully ripen fruits and passed through a fine cotton cloth.

Total soluble solids (TSS) *:* The total soluble solids content of fruit was determined with the help of a hand refractometer, calibrated at $20\,^0$ C.

Total and reducing sugars : The total and reducing sugars contents of the fruit were estimated by titrating against Fehling's A and Fehling's B reagents, using methylene blue as an indicator (AOAC, 1984).

Acidity : Total acidity was determined by titrating against 0.1 N NaoH using phenolpthalin as an indicator (AOAC, 1984).

Ascorbic acid : 2, 6 – dichlorophenol indophenol dye titration method was used to estimate the ascorbic acid content of the fruit (AOAC, 1984).

3.6.6 Tissue analysis : Representative samples of leaf and shoot were collected, oven dried (60^0C) and were ground to pass through a 40 mesh screen and the requisite amount was taken to estimate nitrogen, phosphorus, potassium and total carbohydrate content by using the following methods.

Nitrogen : The total N content (% dry weight basis) was estimated by microkjeldahl method as described by Black (1965).

Phosphorus : Phosphorus was estimated by Vanadomolybdate yellow colour method (Chapman and Pratt, 1961).

Potassium : Potassium content was determined by Flame photometry (Chapman and Pratt, 1961).

Total carbohydrate : Total carbohydrate was estimated by colorimetric method using anthrone as a reagent (Yoshida *et al.,* 1972).

3.6.7 Soil analysis :

Total nitrogen and available nitrogen content of soil : Total soil nitrogen content was determined by modified macro-kjeldahl method

(Jackson, 1967) and available nitrogen was determined by the Alkaline potassium permanganate method (Subbiah and Asija, 1956).

Available phosphorus : Available phosphorus was estimated by Olson method (Jackson, 1967).

Available potassium : Available potassium was determined by Flame photometric method (Jackson, 1967).

Organic carbon : Walkley and Black's rapid titration method was used to determine organic carbon content of the soil (Jackson, 1967).

Soil reaction : Soil pH was measured by glass electrode pH meter in 1: 2.5 soil-water suspension (Jackson, 1967).

*3.6.8 Analysis of micronutrient content in plant and soil samples :*The plant and soil samples were analyzed for micronutrient content from the laboratory of Phosphate and Potash Institute of Canada (PPIC).

*3.6.9 Installation and operation of drip irrigation system :*Necessary arrangements were made for drip irrigation. The quantity of water to be applied per tree was calculated on the basis of the following formula :

Volume of water/tree (V) = E_p x K_c x K_px Area of wetting x Day interval

Here, E_p = Weekly average of evaporation from a class A open pan
        evaporimeter in cm.

Kc = Crop factor (0.80)

Kp = Pan coefficient, for the experimental site it is 0.80.

Area of wetting was considered as 60% of the canopy coverage by tree. The treatments were designed to impose 80% (T_1:0.80E_0), 60% (T_2:0.60E_0), 40% (T_3:0.40E_0), 20% (T_4:0.20E_0) and 0% (T_5:Control) of V.

The operation time of drip system for respective treatments was calculated on the basis of discharge rate of individual dripper (8 l / h) and number of drippers (2) per tree.

*3.6.10 Soil moisture studies :*Weekly variations in soil moisture content at different soil depth was measured by Neutron moisture probe (503 DR

Hydroprobe, CPN, Martinez, California, USA). Initial soil moisture content at different soil depths was determined by thermo-gravimetric method and used for calibration of the instrument.

3.6.11 *Measuring leaf water potential (LWP) and relative water content (RWC)* : Representative leaf samples were collected from the shaded side at 0900 h and sunlit side at 1400 h (only from week 5 for afternoon sampling) for these studies. Leaf water potential was estimated following the "Pressure bomb" technique of Scholander *et al.* (1965). The relative water content of leaf was estimated by using the following formula (Barrs and Weatherley, 1962).

RWC(%)=[(Fresh weight–Dry weight)/(saturated weight–Dry weight)]x100

3.6.12 *Measuring stomatal conductance and net CO_2 assimilation rate* : Net CO_2 assimilation rate, stomatal conductance, leaf temperature and some other parameters were measured weekly for all treatments using a portable Photosynthesis System (CI-310, CID, Inc., USA).

3.6.13 *Estimation of weekly water use (E_t), crop factor ($K_c = E_t/E_0$) and water expense efficiency (WEE):* The water use (E_t) was calculated by water balance method. Crop factor and water use efficiency were estimated using the following formula.

$K_c = E_t/E_p$, E_p is weekly average of evaporation from a class A open pan evaporimeter.

WEE (in terms of marketable yield) = Yield (kg/ha) / E_t (cm).

3.6.14 *Methods of statistical analysis of data* : Data were analyzed for statistical inference following the statistical method for Randomized Block Design (RBD) described by Gomez and Gomez (1983).

3.6.15 *Economics of drip irrigation in litchi* : The benefit : cost ratio of litchi cultivation under drip irrigation system was estimated.

35

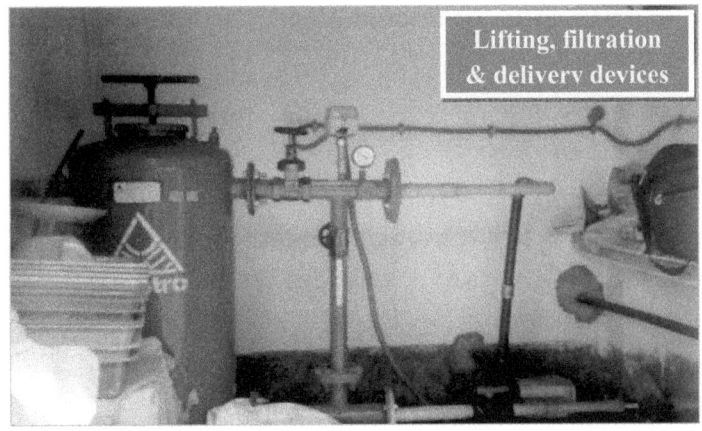

Plate 1: THE DRIP IRRIGATION SYSTEM

OF THE EXPERIMENT

Plate 2: THE DRIP IRRIGATION SYSTEM

OF THE EXPERIMENT

Measuring soil water content using neutron moisture probe

The 'Pressure bomb' accessories

Plate 3 :STUDYING SOIL MOISTURE DYNAMICS AND TREE-WATER RELATIONS IN LITCHI

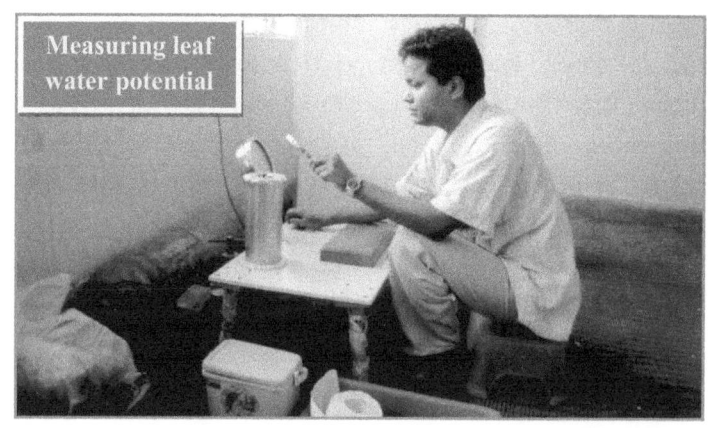

Measuring leaf water potential

Measuring carbon assimilation

Plate 4 : STUDYING TREE-WATER RELATIONS AND CO$_2$ ASSIMILATIONIN BEARING LITCHI TREES

4.0 RESULTS AND DISCUSSION

Experiment I: Relationship between vegetative flushing and leaf N, timing of fertilizer N application.

Among the mineral nutrients, nitrogen (N) plays the most important role in relation to vegetative growth. Vegetative flushing in litchi is known to have a pronounced effect on bearing of the trees. Specific information are, therefore, in demand not only to optimize fertilizer dose but also to standardize the time of fertilizer application, especially N in relation to vegetative flushing and leaf nutrient build up, because these have direct influence on the yield performance of litchi tree.

4.1.1 Monthly shoot growth

October 2001 to September 2002 : Monthly increase in shoot length of the tagged shoots were recorded between 15-17th of each month and the percentage increase over the previous month was calculated. The data recorded for the twelve months period (October 2001 to September 2002) revealed that the increase in growth was higher in the month of October 2001 which then gradually declined in the following months until January which started increasing thereafter and reached maximum in the month of July 2002 (Table 4.1.1). Different treatments showed significant variations in increase of shoot growth in the month of September 2002. The increase in growth (average of five treatments) was recorded maximum (54.18%) in the month of July while it was least (1.18%) in January 2002. Among different treatments, treatment T_3 (two split application of N in July and September) caused the maximum increase of shoot growth followed by T_1, T_5, T_4 and the least in T_2 treatment (application of N in July & March).

October 2002 to September 2003 : The data recorded on monthly shoot growth between October 2002 to September 2003 (Table 4.1.2) revealed that the increase in shoot growth was higher in the month of October 2002 which then gradually declined in the following months until January and again increased further to reach the maximum in the month of July 2003. The different treatments showed significant variations in increase of shoot growth in the months of April, July, August and September 2003. The increase in growth (average of five treatments) was recorded maximum in the month of July 2003 (30.93% increase over the month of June) followed by October 2002 (29.18 % over the month of September) while it was least (0.78 %) in January 2003. Among different treatments, treatment T_5 (control : two split application of N in June and August) caused the maximum increase of shoot growth followed by T_1, T_3 and T_2 and the least in T_4 (three splits: June, September & March).

October 2003 to May 2004 : During October 2003 to May 2004, the growth rate of shoot was higher in the month of October 2003 which then gradually declined in the following months until December 2003 and remained almost static in January 2004 (Table 4.1.3). Shoot growth rate between February and May 2004 ranged between 9.60% and 16.30 % over previous months. Different treatments showed significant variations in increase of shoot growth in the months of October and November in 2003 and April and May in 2004. Among different treatments, treatment T_3 (two splits application of N in July and September) caused the maximum increase of shoot growth followed by T_4, T_2, T_1 and the least in T_5.

The average of monthly shoot growth over three years has been presented in Figure 4.1.1. The growth rate of shoot (average of five treatments) was maximum (42.55%) in the month of July and minimum (0.83%) in January. During flowering and fruit development period (February to May), shoot growth increased between 14.29% and 23.19%. The growth rate was much higher (23.51 to 42.55%) during post harvest vegetative growth period (July to October).

Timing of fertilizer N application, however, showed marked variations in increase of shoot growth (Fig. 4.1.2). Total shoot growth was highest (238.07%) when N was applied in July and September (T_3) but it was lowest when N was applied in June, September and March (T_4). Application of entire dose of N in July (T_1) or at two splits in July and September (T_3) or in June and August (T_5) caused about 10.26% to 19.37% higher shoot growth compared with the application in July and March (T_2) or in June, September and March (T_4).

The effect of N application on shoot growth, however, showed much variations between July and December compared with the fruit growth period (February to May). During February to May, total shoot growth ranged between 72.87% and 76.85%, with less than 4% variations due to treatments. The applied N in March (T_2, T_4), therefore supposed to support the developing fruit rather than supporting vegetative growth. But during July to December, treatments effect showed 11.01% to 18.26% variations in shoot growth and thus indicated that the applied N had pronounced effect on shoot growth during the period (Fig. 4.1.2). Application of total N in June (T_1) or at two splits between fruit harvest and dormancy (T_3, T_5) caused more than 135% shoot growth during July to December however, application of one-third or half of total N during fruit development period (T_4, T_2), suppressed the postharvest shoot growth by about 11% to 18%. Proper timing of N application thus appeared effective in monitoring shoot growth in litchi.

4.1.2 Monthly leaf nitrogen content

September 2001 to September 2002 : The average N content of leaf was 2.10% in September 2001 which gradually declined in the following months and reached 1.68% in May 2002 then increased in the following months (Table 4.1.4). The different treatments showed no significant variations in leaf N content in the first year (September 2001 to September 2002). However, the average N content of leaf varied between 1.70% and 1.80% due to treatments.

October 2002 to September 2003 : In second year, the average N content of leaf was 1.78% in October 2002 which gradually declined in the following months and reached 1.34% in January 2003 and thereafter increased in the following months (Table 4.1.5). Significant variations in leaf N content due to treatments were noted in the months of April, July and September 2003. The average N content of leaf varied between 1.60% and 1.65% due to treatments. Application of one split of fertilizer N after fruit set (T_2, T_3)

caused significantly higher leaf N content in the month of April, 2003. In July, 2003, leaf N content was more than 1.74% due to T_1, T_2 and T_5 while in September, 2003, it was more than 1.67% due to T_1, T_3 and T_5 treatments.

*October 2003 to May 2004 :*The average N content of leaf was recorded 1.62% in October 2003 which gradually declined in the following months and reached 1.34% in January 2004 which increased subsequently in the following months. The different treatments also showed significant variations in leaf N content in the months of October and November in 2003 and March, April and May in 2004 (Table 4.1.6). The average N content of leaf varied between 1.53% and 1.56% due to treatments. Application of total N in two splits i.e., in July and September (T_3) caused significantly higher leaf N content in the months of October and November, 2003. However, application of total N in two splits in July and March (T_2) caused significantly higher leaf N content during fruit growth and development period (March, April and May in 2004). In October 2003, leaf N content was above 1.62% due to T_3, T_4 and T_5 while in November 2003, it was above 1.53% due to T_3 and T_5.

Average of monthly leaf N content over three years showed minimum leaf N (1.46% to 1.50%) in the month of January and maximum (1.70% to 1.76%) in July (Fig. 4.1.3). However, different treatments showed negligible variations (1.65% to 1.66%) in leaf N content. The average data of five treatments showed that the leaf N content increased from 1.48% in January to 1.69% in April, which slightly declined (1.68%) in May and then increased again in June (1.71%). In the following four months (July, August, September and October), it remained almost static at 1.72%, after which it gradually declined until January.

Apart from this general trend, timing of N application caused notable variations in different months. Considering the average of leaf N content between July and November (post harvest vegetative growth period), it was found that the leaf N content during the period varied between 1.69% and 1.73% due to time of N application (Fig. 4.1.4). Leaf N content was maximum (1.73%) due to application of N in July and September (T_3) or in

June and August (T_5) and it was minimum (1.69%) due to T_2 (July and March) and T_4 (June, September and March). During fruit development period (March to May), leaf N content was 1.70% to 1.72% due to T_4 and T_2 compared with 1.64% to 1.67% due to T_1 (entire nitrogen when applied once in July), T_3 and T_5 treatments.

4.1.3 Carbohydrate content and C/N ratio of leaf and shoot

Before flowering (December) in 2001 and at anthesis (March) in 2002 :The data in Table 4.1.7 revealed that the average carbohydrate content of leaf and shoot were 5.45% and 6.30%, respectively in December 2001 which increased at anthesis (6.97% and 7.07%, respectively) in March 2002. Different treatments showed no significant variations in leaf and shoot carbohydrate content before flowering or at anthesis. The average C/N ratio of leaf was 3.04 in December 2001 and 3.99 in March 2002. However, the leaf C/N ratio varied between 2.86 and 3.43 in December 2001 and 3.70 and 4.46 in March 2002 due to treatments.

Before flowering (December) in 2002 and at anthesis (March) in 2003 : The average carbohydrate content of leaf and shoot were 6.96% and 9.45%, respectively in December 2002 which increased at anthesis to 10.79% and 12.93%, respectively in March 2003 (Table 4.1.8). Different treatments showed significant variations in leaf and shoot carbohydrate content before flowering and at anthesis. The average C/N ratio of leaf was 4.79 in December 2002 and 7.08 in March 2003. However, the leaf C/N ratio varied between 4.45 and 5.05 in December 2002 and 5.69 and 8.48 in March 2003 due to treatments. Highest carbohydrate content of leaf and shoot before flowering were recorded due to T_2 i.e., application of N in two splits in July and March.

Before flowering (December) in 2003 and at anthesis (March) in 2004 : The average carbohydrate content of leaf and shoot were 7.20% and 9.01%, respectively in December 2003 which increased at anthesis to 11.11% and 12.92%, respectively in March 2004 (Table 4.1.9). Different treatments showed significant variations in leaf and shoot carbohydrate content before flowering and at anthesis. The average C/N ratio of leaf was 5.14 in

December 2003 and 6.72 in March 2004. The leaf C/N ratio varied between 4.80 and 5.63 in December 2003 and 6.57 and 7.01 in March 2004 due to treatments. Application of total N in two splits in July and March (T_2), caused highest C/N ratio of leaf and shoot before flowering and at anthesis.

The average of carbohydrate content of leaf and shoot before flowering and at anthesis over three years of experimentation has been presented in Figures 4.1.5 and 4.1.6. Before flowering (December), the carbohydrate content of leaf and shoot varied between 6.15% and 6.97% and 7.73% and 9.01%, respectively due to timing of N application (treatments). Maximum carbohydrate content of leaf (6.97%) and shoot (9.01%) in December was recorded when N was applied in two splits in July and March (T_2), while application of N in July and September (T_3) caused minimum (6.17%) carbohydrate content of leaf and the shoot carbohydrate content was found minimum (7.73%) due to T_4 (June, September and March) (Fig.4.1.5).

At anthesis (March), carbohydrate content of leaf and shoot showed variations between 8.48% and 10.29% and 10.50% and 12.03%, respectively due to treatments. Application of total N in two splits in July and March (T_2) caused the maximum carbohydrate content of leaf (10.29%) and shoot (12.03%) but it was minimum when N was applied in July and September (T_3) (Fig. 4.1.6).

4.1.4 Differentiation of shoot

The average number flowering and vegetative shoots in three years period of investigation varied between 29.78% and 40.60% and 59.40% and 70.23%, respectively due to treatments (Table 4.1.10). Application of nitrogen in two splits i.e., after harvest (July) and after fruit set (March) (T_2) caused maximum flowering shoots and minimum vegetative shoots but when applied in two splits in July and September (T_3), there was minimum flowering shoots and maximum vegetative shoots. Timing of fertilizer N application showed significant effect on shoot differentiation in both 2003 and 2004. In the first year (2002) of experimentation, T_5 (June and August) caused maximum flowering shoots (28.60%) but in the following two years

(2003 and 2004) maximum flowering shoots (42.48% and 53.40%, respectively) were recorded due to T_2 (July and March).

The average of number of flowering and vegetative shoots over three years showed that timing of N application caused marked variations in differentiation of shoots (Fig. 4.1.7). Application of N in July and March (T_2) caused maximum number of flowering shoots (40.60%) and minimum number of vegetative shoots (59.40%) compared with 29.78% flowering shoots and 70.23% vegetative shoots due to application of N in July and September (T_3). The number of flowering shoots varied between 29.78% and 34.72% due to T_1 (entire N in July), T_3 and T_5 (June and August) treatments. However, the number of flowering shoots was comparatively higher (35.68% to 40.60%) when half-or one-third of N was applied in March and rest in July (T_2) or in June and September (T_4).

4.1.5 Types of panicle

Production of mixed and pure types of panicles in three years varied between 50.00% and 55.19% and 44.85% and 49.89%, respectively (Table 4.1.10). Pure type panicle was more than 48% in 2002 and 2004 but it was less than 36% in 2003, irrespective of the treatments. Timing of fertilizer N application showed significant effect on types of panicle in the years 2003 and 2004. Maximum number of mixed panicles (51.30%) was recorded due to T_4 (June, September and March) in 2002, while it was maximum (78.00%) due to T_1 (June) in 2003 and T_5 (47.02%) in 2004.

4.1.6 Flowering characters

Length and breadth of panicle: Timing of fertilizer N application showed significant effect on panicle size in the second and third years of experimentation. Maximum size (26.30 cm X 16.03 cm) of panicle was recorded due to T_2 (July and March) and minimum (24.76 cm X 14.41 cm) due to T_3 (July and September) treatment (Table 4.1.11).

*Number of flowers per panicle :*Average number of flowers per panicle varied between 1321.02 and 1480.01 due to different treatments, however, the variations were not significant (Table 4.1.11).

46

Sex ratio and fruit set per panicle : Ratio of staminate and pistillate flowers on a panicle varied between 2.65 : 1 and 3.12 : 1 due to treatments (Table 4.1.12). Fruit set per panicle at harvest showed significant variations in the second and third years due to treatments. Average number of fruit set ranged from 11.80 to 13.72 per panicle. Highest fruit set per panicle was recorded due to T_2 (July and March), closely followed by T_4 (June, September and March) treatments.

The average fruit set per panicle at harvest due to treatments in three years showed that application of N after fruit set (T_2, T_4) caused retention of 13.30 to 13.72 fruits per panicle compared with 11.80 to 12.83 fruit per panicle when N was not applied after fruit set (T_1, T_3 and T_5) (Fig. 4.1.8).

4.1.7 Physical characters of fruit

*Length and diameter of fruit :*Average length and diameter of fruit varied between 4.04 cm and 4.11 cm and 3.31cm and 3.46 cm, respectively due to treatments. The effect of treatments on length and diameter of fruit was not statistically significant (Table 4.1.13). However, application of N in July and March (T_2) caused maximum length (4.11 cm) and diameter (3.46 cm) of fruit compared with 4.04 cm in length and 3.31 cm in diameter of fruit due to application of entire dose of N in July (T_1).

*Fruit weight :*Weight of fruit varied significantly due to treatments in the second and third years of investigation. The average of fruit weight ranged between 20.48g and 21.88 g. Application of one splits of nitrogenous fertilizer (T_2 and T_4) in the month of March caused significantly higher fruit weight compared with other treatments. The fruit weight was recorded maximum (23.43 g) when N was applied in two splits in the months of July and March (T_2) (Table 4.1.13, Fig. 4.1.9).

Aril content : The aril content of fruit varied between 50.50% and 62.61% in different years. However, the time of N application showed considerable variations (56.77% to 59.81%) in aril content of fruit. Trees receiving N application after fruit set (T_2, T_4) produced fruits with higher aril content. Two split applications of total N in July and March (T_2) caused the highest aril content of fruit (59.81%) followed by 58.33% in T_4 i.e., three split applications in June, September and March (Table 4.1.14, Fig. 4.1.9).

4.1.8 Yield

Number of fruits per tree : Application of half- or one-third of total N after fruit set (T_2, T_4) significantly increased the number of fruits per tree in the second and third years of experimentation (Table 4.1.16). The number of fruits (average of three years) varied from 1768.06 to 2032.24 per tree due to treatments. Trees receiving N at 400 g/tree in two splits (July and March) produced the maximum number of fruits (2032.24) compared with 1768.06 in trees receiving N at 400 g/tree in two splits in June and August.

Fruit yield per tree : The yield (kg/tree) varied significantly due to timing of fertilizer N application in the years 2003 and 2004. The average yield over three years period ranged between 35.82 and 43.82 kg/tree due to treatments. Treatment effect on fruit yield showed similar trends as recorded for number of fruits per tree. Tree receiving total N in two splits in July and March (T_2) caused the maximum yield of 43.83 kg/tree compared with 35.82 kg/tree in T_3 treatment (Table 4.1.15, Fig. 4.1.10).

The average yield was 36.85 kg/tree when N was applied following the recommended schedule (two splits, in June and August; T_5) of West Bengal. Highest increase in yield of 17.59% over T_5 was recorded due to T_2 (two splits, in July & March) but application of total N in July and September in two equal splits (T_3) decreased yield by 2.63% over T_5.

4.1.9 Fruit quality

Total soluble solids : Timing of fertilizer N application caused significant variation in total soluble solids (TSS) content of fruit in the second and third years of investigation (Table 4.1.16). Highest TSS content of fruit was recorded due to T_2 treatment which was statistically at par with T_4. However, the average of TSS content of fruit over three years varied between 17.78 (T_3) and 18.58 ^{0}Brix (T_2) due to treatments.

Reducing sugar : Reducing sugar content of fruit ranged between 13.15% and 14.25% in different years due to treatments. However, treatment effect on reducing sugar content of fruit was non-significant (Table 4.1.16).

Total sugars : The total sugars content of fruit over three years showed slight variation due to treatments which varied between 15.21% and 15.74% (Table 4.1.16). However, treatment effect on total sugars content of fruit was significant in 2003 and 2004. Fruit content significantly higher total sugars (15.71% to 15.74%) due to application of fertilizer N after fruit set (T_2, T_4).

*Tritratable acidity :*The fruit acidity content was recorded between 0.43% and 0.63% in different years without any significant effect of treatment. Average of tritratable acidity of fruit over three years was found minimum (0.48%) with T_2 treatment and highest (0.54%) with T_3 treatment (Table 4.1.17).

Ascorbic acid : The ascorbic acid content of fruit (average of three years) was found maximum (32.60 mg ascorbic acid per 100g pulp) by application of N in two splits in the months of July and March (T_2) compared with 18.59 mg/100g pulp when N was applied in June and August (T_5)(Table 4.1.17).

TSS/acid ratio : Timing of fertilizer N application caused variations in TSS/acid ratio of fruit in different years which varied between 27.35 and 44.94 (Table 4.1.17). The average TSS/acid ratio of fruit over three years showed that the ratio was more than 38 due to T_2, T_4 and T_5 and less than 35 due to T_1 and T_3 treatments (Fig. 4.1.10).

4.1.10 Relation between pre-flowering and post-flowering tree nutrition status with flowering, yield and quality of fruit

Correlation matrix of relationship between pre-flowering leaf N content, shoot growth rate, carbohydrate content of leaf and shoot, flowering and yield : The correlation matrix of relationship between pre-flowering leaf N content, shoot growth rate, carbohydrate content of leaf and shoot, flowering and yield of litchi cv. Bombai has been worked out (Table 4.1.18). The data revealed that the pre-flowering (October to December) leaf N content was positively related with shoot growth between October and December but was found negatively related with carbohydrate content of leaf and shoot in December, flowering shoot (%) and yield. The pre-flowering carbohydrate content of leaf and shoot (in December) showed a significant and positive correlation with the number of flowering shoot (%) and yield.

Correlation matrix of relationship between post-flowering leaf N content, carbohydrate content of leaf and shoot, fruit retention, yield and quality (TSS/acid ratio) of fruit : Correlation matrix of relationship between post-flowering leaf N content, carbohydrate content of leaf and shoot, fruit retention, yield and quality (TSS/acid ratio) of fruit of litchi cv. Bombai were also estimated (Table 4.1.19). The leaf N content during fruit growth and development period (April to May) showed significant and positive relation with number of fruits retained and also showed positive correlation with yield and quality (TSS/acid ratio) of fruit. The carbohydrate content of both leaf and shoot in March showed a positive correlation with retention, yield and quality of fruit but it was the shoot carbohydrate content in March that showed significant and positive relation with fruit yield.

4.1.11 DISCUSSION

It is now well established that litchi needs proper nutritional care for sustain production. The nutritional requirement of litchi has also been worked out based on yield record, age or canopy size of a tree (Ghosh and Mitra, 1990, Hasan and Chattapadhyay, 1997, Mitra, 2004b) and conventionally the fertilizer is applied at the onset and end of monsoon in West Bengal.

Litchi showed strong correlation between yield and vegetative growth, which is influenced by leaf N content (Koen *et al.*, 1981, Koen and Smart, 1982). It is therefore, possible that the time of fertilizer application (especially N) and leaf N content could have pronounced effect on the bearing of litchi trees.

In the present experiment, the trees were fertilized with N, P_2O_5 and K_2O @ 400, 200 and 400 g/tree/year, respectively. The P_2O_5 and K_2O were applied in two splits i.e., in July and March, while the time of N application constituted the treatments, viz., i) after harvest (July)(T_1), ii) in two splits, after harvest (July) and after fruit set (March) (T_2) iii) in two splits, after harvest (July) and in autumn (September) (T_3) iv) in three splits, June, September and March. (T_4) v) conventional method (as control) in two splits, onset and the end of monsoon (T_5). Regarding the crop growth periods, total N was applied between fruit harvest (June) and dormancy (December) in case of T_1 (entire N applied in July), T_3 (in two splits i.e., July and September) and T_5 (in two splits i.e., July and August) treatments. Application of N was done at vegetative growth period as well as fruit growth period in case of T_2 (in two splits i.e., July and March) and T_4 (in three splits i.e., July, September and March) treatments.

The results revealed that the timing of N application caused marked variations in increase of shoot growth. Total shoot growth was maximum (238.07%) when N was applied in July and September (T_3) but it was lowest when N was applied in three splits in June, September and March (T_4). Application of entire dose of N in July (T_1) or at two splits in July and September (T_3) or in June and August (T_5) caused about 10.26% to 19.37% higher shoot growth compared with the application in July and March (T_2) or in June, September and March (T_4). Total N when applied in July (T_1) or in two splits between fruit harvest and dormancy (T_3, T_5) caused more than 135% shoot growth during July to December, however application of one-third or half of total N during the fruit development period (T_4, T_2), reduced the postharvest shoot growth by about 11% to 18%. The leaf N content during post harvest vegetative growth period (between July and November) varied between 1.69% and 1.73% due to time of N application. The average leaf N content during the post harvest vegetative growth period was higher (1.73%) due to application of N in July and September (T_3) or in June and August (T_5) and it was lower (1.69%) when N was applied in July and March (T_2) and in June, September and March (T_4). The time of N application thus indicated a direct effect on leaf N content and shoot growth (Menzel and Simpson, 1986c, Menzel et al., 1994).

During February to May, total shoot growth ranged between 72.87% and 76.85%, with less than 4% variations due to treatments. While the leaf N content during fruit development period (March to May) was 1.70% to 1.72% due to T_4 and T_2 treatments compared with 1.64% to 1.67% due to T_1 (entire nitrogen in July), T_3 and T_5 treatments. The applied N in March (T_2, T_4), therefore supposed to support the developing fruit rather than supporting vegetative growth because the developing fruits represented the stronger demand pool for nutrients (Hieke et al., 2002a).

Maximum carbohydrate content of leaf (6.97%) and shoot (9.01%) in December was recorded when N was applied in two splits in July and March (T_2), while application of N in July and September (T_3) caused minimum (6.17%) carbohydrate content of leaf and the shoot carbohydrate content was found minimum (7.73%) due to T_4 (June, September and March) treatment. The T_2 treatment produced the maximum number of flowering shoots (40.60%) and minimum number of vegetative shoots (59.40%) compared with 29.78% flowering shoots and 70.23% vegetative shoots due to T_3 treatment. The buildup of higher carbohydrate reserves prior to flowering was found essential for flowering compared with lower reserves in vegetative tree (Menzel et al., 1995b).

At anthesis (March), the maximum carbohydrate content of leaf (10.29%) and shoot (12.03%) were recorded due to T_2 treatment but it was minimum when N was applied in July and September (T_3). The number of fruits retained per panicle at harvest due to treatments showed that application of N after fruit set (T_2, T_4) caused retention of 13.30 to 13.72 fruits per panicle compared with 11.80 to 12.83 fruits per panicle when N was not applied after fruit set (T_1, T_3 and T_5). Although fruit development principally depended on current assimilation, the use of reserved carbohydrate by the developing fruit could not be over ruled (Roe et al., 1997, Hieke et al., 2002b).

Application of one splits of nitrogenous fertilizer (T_2 and T_4) in the month of March caused significantly higher fruit weight compared with other treatments. The fruit weight was recorded maximum (23.43 g) when N was applied in two splits in the months of July and March (T_2). Trees receiving N application after fruit set (T_2, T_4) produced fruits with higher aril content. Two splits application of total N in July and March (T_2) caused the highest aril content of fruit (59.81%) followed by 58.33% in T_4 i.e., application in June, September and March. Fertilization with N after fruit set thus appeared beneficial probably because the N is an important constituent of protoplasm and increases the size and total number of cells (Njoku, 1957).

Trees receiving N at 400 g/tree in two splits (July and March) produced the maximum number of fruits (2032.24/tree) compared with 1768.06 in trees receiving N at 400 g/tree in June and August (T_3). The yield (kg/tree) varied significantly due to timing of fertilizer N application. Tree receiving total N in two splits at July and March (T_2) produced highest yield of 43.83 kg/tree compared with 35.82 kg/tree in T_3 treatment. The timing of fertilizer N application caused significant variation in total soluble solids (TSS) content of fruit in the second and third years of investigation. Highest TSS content of fruit was recorded due to T_2 which was statistically at par with T_4 treatment. Significantly higher total sugars content (15.71% to 15.74%) of fruit was recorded due to application of fertilizer N after fruit set (T_2, T_4). The tritratable acidity of fruit was minimum (0.48%) with T_2 treatment and highest (0.54%) with T_3 treatment. The TSS/acid ratio of fruit was more than 38 due to T_2, T_4 and T_5 and less than 35 due to T_1 and T_3 treatments. The developing fruits with a large nutrients demand pool are the most important sink of a bearing tree. The uptake of nutrients and accumulation in litchi fruits continued throughout the fruit growth period (Menzel et al., 1988) and fertilization after flowering, therefore supposed to increase the size, weight and yield of fruit and improved the fruit quality (sugar/acid ratio) and reduced malformed fruits (Koen et al., 1981). Application of one split of fertilizer N after fruit set resulted in better growth of the developing fruits, higher fruit retention and yield and quality of fruit probably because nitrogen, being the major constituents of many compounds of great physiological importance in metabolisms, associated in synthesis of protein, chlorophyll, enzymes, hormones and vitamins (Bloom, 1992).

The correlation matrix of relationship between pre-flowering leaf N content, shoot growth rate, carbohydrate content of leaf and shoot, flowering and yield of litchi cv. Bombai revealed that the pre-flowering (October to December) leaf N content had a positive relation with shoot growth rate during October to December but showed a negative relation with carbohydrate content of leaf and shoot in December, flowering shoot (%) and yield. Studies by other workers also showed that although there was a strong correlation between yield and leaf N upto 14.7 mg N/g, very high rates of N depressed yield, presumably because of excessive vegetative flushing (Langenegger, 1975, Koen *et al.*, 1981, Koen and Smart, 1982). The pre-flowering carbohydrate content of leaf and shoot (in December) showed significant and positive correlation with the number of flowering shoot (%) and yield.

The leaf N content during fruit growth and development period (April to May) showed significant and positive correlation with fruit retention and also positive relationships with yield and quality (TSS/acid ratio) of fruit. The carbohydrate content of both leaf and shoot in March had also positive correlation with retention, yield and quality of fruit but it was the shoot carbohydrate content in March that showed significant and positive relation with fruit yield. Zhuang *et al.*(1988) showed that poor fruiting in litchi was associated with low levels of nitrate –N (<960 µg/g), P (< 0.5 mg/g) and K (< 11.6 mg/g) in the leaf prior to and after fruit set. Hauang *et al.* (1998) also developed a relationship between leaf N content and yield, and showed that fruit yield should be guaranteed to be more than 90% of the maximum when leaf N content remained in the range of 1.5% to 1.9 % at anthesis.

Table 4.1.1(a) : Increase of shoot growth in different months (Oct., 2001 – Sept., 2002).

Treatment (Time of N application)	% of increase over previous month $^{\varphi}$						Aver-age
	Oct. 01	Nov. 01	Dec. 01	Jan. 02	Feb. 02	Mar. 02	
T_1 (Entire dose in July)	35.98 (36.80) **	16.86 (4.09) *	4.18 (2.11) *	1.50 (1.41) *	40.56 (39.53) **	29.68 (32.73) **	24.68
T_2(Two splits: July & March)	19.77 (26.33)	9.80 (3.12)	6.17 (2.58)	0.99 (1.22)	39.11 (38.66)	19.56 (24.27)	22.14
T_3(Two splits: July & Sept.)	38.32 (38.22)	16.81 (4.08)	10.16 (3.26)	1.20 (1.30)	28.62 (32.31)	35.26 (35.99)	25.87
T_4 (Three splits: June, Sept. & March)	29.35 (32.76)	20.30 (4.48)	8.27 (2.93)	0.47 (0.98)	32.48 (34.65)	33.78 (34.78)	22.94
T_5(Control: June & Aug., Normal practice in W.B.)	27.84 (31.78)	14.87 (3.82)	4.97 (2.33)	1.72 (1.49)	39.41 (38.79)	28.94 (31.72)	23.53
Total	151.3	78.64	33.75	5.88	180.2	147.22	-
Average	30.25	15.73	6.75	1.18	36.04	29.44	-
SEm (±)	1.349	0.240	0.194	0.061	2.009	6.037	-
C.D. at 5%	4.154	0.739	0.596	0.189	NS	NS	-

* Square root transformed values,
** Angular transformed values,
NS=Non-significant
$^{\varphi}$ Initial shoot length (average of treatments) : 9.70 cm. (September, 2001).

Table 4.1.1(b) : Increase of shoot growth in different months (Oct., 2001 – Sept., 2002).

Treatment (Time of N application)	% of increase over previous month φ						Total	Average
	Apr. 02	May, 02	June, 02	July, 02	Aug. 02	Sept. 02		
T_1 (Entire dose in July)	24.50 (27.32)**	11.00 (18.08)**	11.48 (22.22)**	57.17	33.10 (34.52)**	30.20 (33.27)**	296.2	24.68
T_2(Two splits: July & March)	21.35 (26.65)	34.02 (35.18)	17.50 (24.32)	47.46	21.84 (27.32)	28.11 (32.00)	265.7	22.14
T_3(Two splits: July & Sept.)	24.25 (29.29)	33.15 (34.33)	14.89 (21.45)	52.72	21.24 (25.48)	33.65 (35.43)	310.3	25.87
T_4 (Three splits: June, Sept. & March)	22.30 (28.05)	26.00 (30.08)	14.83 (22.13)	51.98	14.75 (22.11)	20.72 (27.05)	275.2	22.94
T_5(Control: June & Aug., Normal practice in W.B.)	24.42 (29.25)	22.35 (27.78)	15.63 (22.15)	61.56	13.07 (20.49)	27.53 (31.62)	282.3	23.53
Total	116.8	86.52	74.33	270.9	104.0	140.21	-	-
Average	23.36	17.30	14.87	54.18	20.80	28.04	-	-
SEm (±)	4.274	4.332	3.391	4.233	3.752	0.817	-	-
C.D. at 5%	NS	NS	NS	NS	NS	2.515	-	-

* Square root transformed values, ** Angular transformed values, **NS**=Non-significant,
φ Initial shoot length (average of treatments) : 9.70 cm. (September, 2001).

56

Table 4.1.2 (a): Increase of shoot growth in different months (Oct., 2002 – Sept., 2003).

Treatment (Time of N application)	% of increase over previous month						Average
	Oct. 02	Nov. 02	Dec. 02	Jan. 03	Feb. 03	Mar. 03	
T_1 (Entire dose in July)	27.96 (16.02) **	12.17 (3.42) *	1.36 (2.52) *	0.47 (0.96) *	16.24 (4.01) *	13.51 (3.65) *	16.39
T_2 (Two splits: July & March)	26.80 (15.35)	12.08 (3.45)	1.67 (2.58)	0.59 (1.02)	17.33 (4.07)	12.81 (3.55)	15.95
T_3 (Two splits: July & Sept.)	29.97 (17.16)	14.35 (3.76)	1.96 (2.63)	0.62 (1.05)	17.78 (4.21)	11.78 (3.42)	16.36
T_4 (Three splits: June, Sept. & March)	30.94 (17.76)	11.80 (3.42)	1.68 (2.59)	0.63 (1.04)	17.08 (4.12)	12.51 (3.52)	15.51
T_5 (Control: June & Aug., Normal practice in W.B.)	30.21 (17.30)	13.52 (3.67)	1.69 (2.58)	0.82 (1.14)	17.21 (4.13)	13.49 (3.65)	16.79
Total	145.9	63.92	8.36	3.13	85.64	64.10	-
Average	29.18	12.78	1.67	0.78	17.13	12.82	-
SEm (±)	1.934	0.290	0.092	0.112	0.312	0.276	-
C.D. at 5%	NS	NS	NS	NS	NS	NS	-

* Square root transformed values,
** Angular transformed values,
NS= Non-significant

Table 4.1.2 (b) : Increase of shoot growth in different months (Oct., 2002 – Sept., 2003).

Treatment (Time of N application)	% of increase over previous month						Total	Aver-age
	Apr. 03	May, 03	June, 03	July, 03	Aug. 03	Sept. 03		
T₁ (Entire dose in July)	10.36 (3.22) *	7.43 (2.73) *	22.45 (4.74) *	33.22 (19.04) **	31.35 (17.96) **	20.11 (4.48) *	196.9	16.39
T₂ (Two splits: July & March)	15.72 (3.96)	9.62 (3.10)	19.69 (4.44)	30.50 (17.48)	27.18 (15.57)	17.36 (4.17)	191.3	15.95
T₃ (Two splits: July & Sept.)	12.18 (3.49)	8.81 (2.97)	20.73 (4.55)	28.72 (16.46)	30.18 (17.29)	19.20 (4.38)	196.3	16.36
T₄ (Three splits: June, Sept. & March)	12.98 (3.60)	8.78 (2.96)	20.46 (4.52)	29.28 (16.78)	25.49 (14.61)	14.45 (3.80)	186.1	15.51
T₅ (Control: June & Aug., Normal practice in W.B.)	13.45 (3.67)	6.55 (2.56)	21.32 (4.62)	32.94 (18.88)	26.52 (15.20)	23.81 (4.88)	201.5	16.79
Total	64.69	41.19	104.6	154.7	140.7	94.93	-	-
Average	12.94	8.24	20.93	30.93	28.18	18.99	-	-
SEm (±)	0.591	0.911	1.623	0.614	0.323	0.983	-	-
C.D. at 5%	2.181	NS	NS	1.891	0.994	3.026	-	-

* Square root transformed values,

** Angular transformed values,

NS=Non-significant

58

Table 4.1.3 : Increase of shoot growth in different months (Oct., 2003 – May, 2004).

Treatment (Time of N application)	% of increase over last month								Ave-rage
	Oct. 03	Nov. 03	Dec. 03	Jan. 04	Feb. 04	Mar. 04	Apr. 04	May. 04	
T_1 (Entire dose in July)	22.25 (4.72)*	12.60 (3.55)*	1.66 (1.47)*	0.63 (1.06)*	15.72 (3.96)*	17.01 (4.12)*	16.08 (4.01)*	9.61 (3.10)*	11.9
T_2(Two splits: July & March)	21.34 (4.62)	11.80 (3.43)	0.92 (1.19)	0.91 (1.19)	17.63 (4.20)	16.43 (4.05)	17.15 (4.14)	9.82 (3.13)	12.0
T_3(Two splits: July & Sept.)	25.72 (5.07)	15.57 (3.95)	2.11 (1.61)	0.76 (1.12)	16.48 (4.06)	15.41 (3.93)	15.51 (3.94)	8.70 (2.95)	12.5
T_4 (Three splits: June, Sept. & March)	24.68 (4.97)	12.50 (3.67)	1.01 (1.23)	0.55 (1.02)	16.25 (3.90)	15.38 (3.92)	16.75 (4.09)	9.50 (3.08)	12.1
T_5(Control : June & Aug., Normal practice in W.B.)	19.66 (4.43)	14.00 (3.74)	1.51 (1.42)	0.61 (1.05)	15.66 (3.96)	17.26 (4.15)	15.92 (3.99)	8.98 (3.00)	11.7
Total	110.65	64.47	7.21	3.46	79.74	81.49	81.41	48.01	-
Average	22.13	12.89	1.44	0.69	15.95	16.30	16.28	9.60	-
SEm (±)	0.979	0.946	1.721	0.883	0.922	2.114	0.501	0.391	-
C.D. at 5%	3.015	2.913	NS	NS	NS	NS	1.546	1.203	-

* Square root transformed values,

NS=Non-significant

Table 4.1.4 (a) : Nitrogen content of leaf in different months (Sept., 2001 – Sept., 2002).

Treatment (Time of N application)	Monthly nitrogen content of leaf (% dry weight)							Average
	Sept. 01	Oct. 01	Nov. 01	Dec. 01	Jan. 02	Feb. 02	Mar. 02	
T_1 (Entire dose in July)	2.12	1.91	1.86	1.80	1.78	1.75	1.73	1.73
T_2(Two splits: July & March)	2.10	1.86	1.86	1.78	1.75	1.72	1.73	1.72
T_3(Two splits: July & Sept.)	2.08	1.92	1.90	1.84	1.80	1.78	1.76	1.71
T_4 (Three splits: June, Sept. & March)	2.10	1.88	1.88	1.82	1.78	1.76	1.79	1.70
T_5(Control: June & Aug., Normal practice in W.B.)	2.10	1.82	1.80	1.75	1.75	1.74	1.74	1.80
Average	2.10	1.88	1.86	1.80	1.77	1.75	1.75	-
SEm ($\pm$)	-	0.121	0.088	0.133	0.119	0.135	0.011	-
C.D. at 5%	-	NS	NS	NS	NS	NS	NS	-

NS=Non-significant

Table 4.1.4 (b) : Nitrogen content of leaf in different months (Sept., 2001 – Sept., 2002).

Treatment (Time of N application)	Monthly nitrogen content of leaf (% dry weight)						Aver-age
	Apr. 02	May, 02	June, 02	July, 02	Aug. 02	Sept. 02	
T_1 (Entire dose in July)	1.61	1.65	1.72	1.79	1.77	1.66	1.73
T_2(Two splits: July & March)	1.76	1.77	1.77	1.76	1.83	1.64	1.72
T_3(Two splits: July & Sept.)	1.66	1.67	1.63	1.54	1.68	1.65	1.71
T_4 (Three splits: June, Sept. & March)	1.87	1.66	1.64	1.73	1.71	1.60	1.70
T_5(Control: June & Aug., Normal practice in W.B.)	1.71	1.67	1.78	1.91	1.65	1.82	1.80
Average	1.72	1.68	1.71	1.75	1.73	1.67	-
SEm (±)	0.074	0.094	0.010	1.962	0.074	0.324	-
C.D. at 5%	NS	NS	NS	NS	NS	NS	-

NS=Non-significant

Table 4.1.5 (a) : Nitrogen content of leaf in different months (Oct., 2002 – Sept., 2003).

Treatment (Time of N application)	Monthly nitrogen content of leaf (% dry weight)						Aver-age
	Oct. 02	Nov. 02	Dec. 02	Jan. 03	Feb. 03	Mar. 03	
T_1 (Entire dose in July)	1.78	1.60	1.39	1.33	1.40	1.56	1.61
T_2(Two splits: July & March)	1.70	1.76	1.39	1.27	1.38	1.69	1.64
T_3(Two splits: July & Sept.)	1.73	1.81	1.59	1.37	1.30	1.43	1.61
T_4 (Three splits: June, Sept. & March)	1.81	1.60	1.50	1.32	1.39	1.39	1.60
T_5(Control: June & Aug., Normal practice in W.B.)	1.89	1.73	1.41	1.41	1.47	1.57	1.65
Average	1.78	1.70	1.46	1.34	1.39	1.53	-
SEm (±)	0.111	0.084	0.091	0.115	0.112	0.099	-
C.D. at 5%	NS	NS	NS	NS	NS	NS	-

NS=Non-significant

Table 4.1.5 (b) : Nitrogen content of leaf in different months (Oct., 2002 – Sept., 2003).

Treatment (Time of N application)	Monthly nitrogen content of leaf (% dry weight)						Aver-age
	Apr. 03	May, 03	June, 03	July, 03	Aug.03	Sept. 03	
T_1 (Entire dose in July)	1.62	1.66	1.69	1.76	1.74	1.68	1.61
T_2(Two splits: July & March)	1.70	1.72	1.71	1.75	1.74	1.64	1.64
T_3(Two splits: July & Sept.)	1.69	1.69	1.71	1.73	1.73	1.68	1.61
T_4 (Three splits: June, Sept. & March)	1.71	1.70	1.72	1.71	1.69	1.66	1.60
T_5(Control: June & Aug., Normal practice in W.B.)	1.67	1.68	1.67	1.75	1.70	1.69	1.65
Average	1.68	1.69	1.70	1.74	1.72	1.67	-
SEm ($\pm$)	0.009	0.0216	0.035	0.012	1.012	0.013	-
C.D. at 5%	0.028	NS	NS	0.036	NS	0.041	-

NS=Non-significant

Table 4.1.6 : Nitrogen content of leaf in different months
(Oct., 2003 – May, 2004).

Treatment (Time of N application)	Monthly nitrogen content of leaf (% dry weight)								Aver -age
	Oct. 03	Nov. 03	Dec. 03	Jan. 04	Feb. 04	Mar. 04	Apr. 04	May, 04	
T_1 (Entire dose in July)	1.54	1.53	1.40	1.34	1.50	1.64	1.66	1.63	1.53
T_2(Two splits: July & March)	1.56	1.51	1.39	1.35	1.52	1.68	1.72	1.68	1.55
T_3(Two splits: July & Sept.)	1.68	1.56	1.42	1.33	1.51	1.65	1.68	1.65	1.56
T_4 (Three splits: June, Sept. & March)	1.68	1.53	1.40	1.32	1.50	1.64	1.70	1.65	1.55
T_5(Control: June & Aug., Normal practice in W.B.)	1.63	1.54	1.39	1.34	1.51	1.66	1.67	1.62	1.54
Average	1.62	1.53	1.04	1.34	1.51	1.65	1.69	1.65	-
SEm (±)	0.032	0.011	2.0.13	1.109	0.938	0.040	0.009	0.033	-
C.D. at 5%	0.101	0.036	NS	NS	NS	0.123	0.029	0.102	-

NS=Non-significant

Table 4.1.7 : Carbohydrate content and C/N ratio of leaf and shoot before flowering (Dec., 2001) and at anthesis (March, 2002).

Treatment (Time of N application)	Carbohydrate content (% dry wt.)				C/N ratio			
	December, 01		March, 02		December, 01		March, 02	
	Leaf	Shoot	Leaf	Shoot	Leaf	Shoot	Leaf	Shoot
T_1 (Entire dose in July)	5.14	6.20	6.50	6.63	2.86	4.60	3.76	5.06
T_2(Two splits: July & March)	6.01	6.78	7.72	7.92	3.43	5.18	4.46	6.55
T_3(Two splits: July & Sept.)	5.21	6.10	6.81	6.91	2.83	4.36	3.87	5.04
T_4 (Three splits: June, Sept. & March)	5.36	6.15	7.20	7.10	2.95	4.92	4.14	5.68
T_5(Control: June & Aug., Normal practice in W.B.)	5.46	6.25	6.62	6.81	3.12	4.81	3.70	4.97
Average	5.45	6.30	6.97	7.07	3.04	4.77	3.99	5.46
SEm ($\pm$)	2.106	1.922	2.365	0.983	-	-	-	-
C.D. at 5%	NS	NS	NS	NS	-	-	-	-

NS= Non-significant

Table 4.1.8 : Carbohydrate content and C/N ratio of leaf and shoot leaf before flowering (Dec., 2002) and at anthesis (March, 2003).

Treatment (Time of N application)	Carbohydrate content (% dry wt.)				C/N ratio			
	December, 02		March, 03		December, 02		March, 03	
	Leaf	Shoot	Leaf	Shoot	Leaf	Shoot	Leaf	Shoot
T_1 (Entire dose in July)	6.23	8.75	11.07	11.91	4.48	7.29	7.10	9.02
T_2(Two splits: July & March)	7.07	10.96	11.37	14.53	5.05	8.56	8.48	11.18
T_3(Two splits: July & Sept.)	6.44	9.69	8.13	12.08	4.45	8.01	5.69	9.66
T_4 (Three splits: June, Sept. & March)	7.02	9.80	11.78	13.27	4.64	6.09	7.38	10.98
T_5(Control: June & Aug., Normal practice in W.B.)	6.96	8.04	11.59	12.85	4.63	7.78	6.73	9.52
Average	6.96	9.45	10.79	12.93	4.79	7.55	7.08	10.07
SEm (±)	0.369	0.629	0.715	0.345	-	-	-	-
C.D. at 5%	1.137	1.937	2.202	1.063	-	-	-	-

NS= Non-significant

Table 4.1.9 : Carbohydrate content and C/N ratio of leaf and shoot before flowering (Dec., 2003) and at anthesis (March, 2004).

Treatment (Time of N application)	Carbohydrate content (% dry wt.)				C/N ratio			
	December, 03		March, 04		December, 03		March, 04	
	Leaf	Shoot	Leaf	Shoot	Leaf	Shoot	Leaf	Shoot
T_1 (Entire dose in July)	7.15	8.80	11.06	12.97	5.11	7.09	6.74	9.90
T_2(Two splits: July & March)	7.82	9.30	11.78	13.65	5.63	7.75	7.01	10.11
T_3(Two splits: July & Sept.)	6.81	8.86	10.50	12.10	4.80	6.66	6.60	9.45
T_4 (Three splits: June, Sept. & March)	7.25	9.10	11.32	13.01	5.22	7.15	6.90	10.01
T_5(Control: June & Aug., Normal practice in W.B.)	6.95	9.01	10.91	12.86	4.96	7.00	6.57	9.60
Average	7.20	9.01	11.11	12.92	5.14	7.13	6.72	9.81
SEm (±)	0.129	0.069	0.201	0.192	-	-	-	-
C.D. at 5%	0.418	0.213	0.620	0.592	-	-	-	-

NS= Non-significant

Table 4.1.10 (a) : Shoot differentiation and types of panicles due to different time of nitrogen application.

Treatment (Time of N application)	Differentiation of shoot							
	Flowering shoot (%)				Vegetative shoot (%)			
	2002	2003	2004	Aver -age	2002	2003	2004	Aver -age
T₁(Entire dose in July)	26.70 (5.17) *	31.27	46.18	34.72	73.28 (8.56) *	68.75	53.80	65.28
T₂ (Two splits : July & March)	25.95 (5.09)	42.48	53.40	40.60	74.09 (8.61)	57.50	46.62	59.40
T₃ (Two splits : July & Sept.)	24.50 (4.95)	28.12	36.73	29.78	75.50 (8.69)	71.89	63.30	70.23
T₄(Three splits : June, Sept. & March)	26.90 (5.19)	31.73	48.40	35.68	72.90 (8.54)	70.20	51.64	64.91
T₅(Control : June & August, Normal practice in WB)	28.60 (5.35)	29.97	39.80	32.79	71.42 (8.45)	73.00	60.21	68.21
SEm(±)	2.613	1.091	0.757	-	1.812	1.732	1.067	-
C.D. at 5%	NS	3.360	2.331	-	NS	5.333	3.287	-

* Angular transformed values in parenthesis. NS= Non-significant

Table 4.1.10 (b) : Shoot differentiation and types of panicles due to different time of nitrogen application.

Treatment (Time of N application)	Types of panicles							
	Mixed type (%)				Pure type (%)			
	2002	2003	2004	Aver-age	2002	2003	2004	Aver-age
T$_1$(Entire dose in July)	42.70	78.00	42.25	54.32	58.30	22.06	57.76	46.04
T$_2$ (Two splits : July & March)	47.00	64.50	38.74	50.00	52.93	35.50	61.25	49.89
T$_3$ (Two splits : July & Sept.)	51.05	70.05	43.75	54.95	49.50	30.00	56.25	45.25
T$_4$(Three splits : June, Sept. & March)	51.30	65.00	40.50	52.27	48.60	35.03	59.51	47.71
T$_5$(Control : June & August, Normal practice in WB)	44.30	74.25	47.02	55.19	55.80	25.74	53.00	44.85
SEm($\pm$)	4.304	2.679	1.301	-	3.561	0.669	0.931	-
C.D. at 5%	NS	8.253	4.006	-	NS	2.061	2.869	-

* Angular transformed values in parenthesis. NS= Non-significant

Table 4.1.11 : Variation in panicle size and number of flowers per panicle due to different timing of nitrogen application.

Treatment (Time of N application)	Length of panicle (cm)				Flowers / panicle			
	2002	2003	2004	Aver-age	2002	2003	2004	Aver-age
T₁(Entire dose in July)	31.56	15.15	28.10	24.94	1655.1	790.0	1518.0	1321.02
T₂ (Two splits : July & March)	31.27	18.50	29.13	26.30	1589.3	1172.9	1678.1	1480.01
T₃ (Two splits : July & Sept.)	29.18	17.86	27.25	24.76	1721.2	941.3	1601.6	1421.32
T₄(Three splits : June, Sept. & March)	30.75	17.19	28.64	25.53	1777.0	1022.0	1590.1	1463.02
T₅(Control : June & August, Normal practice in WB)	32.10	16.04	26.75	24.96	1894.4	722.7	1648.4	1421.70
SEm(±)	2.512	0.552	0.615	-	8.970	108.4	641.7	-
C.D. at 5%	NS	1.700	1.893	-	NS	NS	NS	-

NS= Non-significant

Table 4.1.12 : Sex ratio of flowers and final fruit set per panicle due to time of nitrogen application.

Treatment (Time of N application)	Sex ratio (staminate : pistillate)				Fruits set per panicle			
	2002	2003	2004	Aver-age	2002	2003	2004	Aver-age
T_1(Entire dose in July)	3.22 : 1	2.50:1	2.70:1	2.81:1	7.96	13.61	16.91	12.83
T_2 (Two splits : July & March)	2.96 : 1	2.41:1	2.57:1	2.65:1	8.21	14.26	18.68	13.72
T_3 (Two splits : July & Sept.)	3.67 : 1	2.51:1	3.17:1	3.12:1	7.80	12.75	17.77	12.77
T_4(Three splits : June, Sept. & March)	3.30 : 1	3.11:1	2.70:1	3.04:1	8.11	13.92	17.89	13.30
T_5(Control : June & August, Normal practice in WB)	3.46 : 1	2.51:1	3.00:1	2.99:1	8.10	10.65	16.65	11.8
SEm(±)	-	-	-	-	0.880	0.461	0.330	-
C.D. at 5%	-	-	-	-	NS	1.421	1.018	-

NS= Non-significant

Table 4.1.13 (a) : Variations in length and diameter of fruit due to different timimg of nitrogen application.

Treatment	Fruit length (cm)				Fruit diameter (cm)			
(Time of N application)	2002	2003	2004	Aver-age	2002	2003	2004	Aver-age
T_1(Entire dose in July)	4.09	3.99	4.03	4.04	3.09	3.33	3.52	3.31
T_2 (Two splits : July & March)	4.00	4.15	4.17	4.11	3.15	3.59	3.64	3.46
T_3 (Two splits : July & Sept.)	4.01	4.09	4.08	4.06	3.01	3.47	3.51	3.33
T_4(Three splits : June, Sept. & March)	4.01	4.14	4.11	4.09	3.10	3.49	3.53	3.37
T_5(Control : June & August, Normal practice in WB)	3.91	4.12	4.10	4.04	3.08	3.42	3.55	3.35
SEm(±)	0.076	0.167	0.253	-	0.084	0.211	0.314	-
C.D. at 5%	N.S.	NS	NS	-	N.S.	NS	NS	-

NS= Non-significant

Table 4.1.13 (b) : Variations in weight of fruit and peel due to different timimg of nitrogen application.

Treatment (Time of N application)	Fruit weight (g)				Peel weight (g)			
	2002	2003	2004	Average	2002	2003	2004	Average
T_1(Entire dose in July)	17.92	21.87	21.65	20.48	4.60	4.63	4.46	4.56
T_2 (Two splits : July & March)	18.85	23.35	23.43	21.88	4.12	4.67	4.67	4.89
T_3 (Two splits : July & Sept.)	18.12	21.47	21.89	20.49	4.22	4.71	4.68	4.54
T_4(Three splits : June, Sept. & March)	18.73	22.27	22.83	21.28	4.62	4.79	4.77	4.73
T_5(Control : June & August, Normal practice in WB)	17.99	21.84	22.14	20.66	4.17	4.81	4.65	4.54
SEm($\pm$)	0.785	0.367	0.338	-	0.227	0.093	0.087	-
C.D. at 5%	N.S.	1.131	1.041	-	N.S.	NS	NS	-

NS= Non-significant

Table 4.1.14 : Variations in seed weight and aril content due to different timing of nitrogen application.

Treatment (Time of N application)	Seed weight (g)				Aril content (%)			
	2002	2003	2004	Average	2002	2003	2004	Average
T_1(Entire dose in July)	4.27	4.27	3.99	4.18	50.50	59.30	60.50	56.77
T_2 (Two splits : July & March)	3.70	4.16	4.09	3.98	54.64	62.18	62.61	59.81
T_3 (Two splits : July & Sept.)	4.00	4.41	3.91	4.11	52.70	57.52	60.94	57.05
T_4(Three splits : June, Sept. & March)	4.12	4.33	3.94	4.13	53.65	59.05	62.29	58.33
T_5(Control : June & August, Normal practice in WB)	3.95	4.38	3.95	4.09	53.08	57.92	60.74	57.25
SEm($\pm$)	0.234	0.677	0.108	-	-	-	-	-
C.D. at 5%	N.S.	NS	NS	-	-	-	-	-

NS= Non-significant

Table 4.1.15 (a): Yield of fruit due to different time of nitrogen application.

Treatment	Number of fruits/tree				Yield (Kg/tree)			
(Time of N application)	2002	2003	2004	Average	2002	2003	2004	Average
T_1(Entire dose in July)	1392.0	1927.8	2265.6	1861.8	23.31	42.16	49.04	38.17
T_2 (Two splits : July & March)	1661.1	2134.1	2301.4	2032.2	26.45	49.83	53.57	43.83
T_3 (Two splits : July & Sept.)	1400.9	1761.1	2146.6	1769.5	22.67	37.81	46.99	35.82
T_4(Three splits : June, Sept. & March)	1373.5	1857.7	2153.3	1794.8	24.18	41.37	49.17	38.24
T_5(Control : June & August, Normal practice in WB)	1326.6	1798.1	2179.5	1768.1	23.03	39.27	48.25	36.85
SEm(±)	350.88	268.81	185.13	-	5.891	2.237	1.152	-
C.D. at 5%	N.S.	NS	NS	-	N.S.	6.891	3.548	-

NS= Non-significant

Table 4.1.15 (b): Increase/decrease (%) of yield over control and total soluble solid content of fruit due to different time of nitrogen application.

Treatment (Time of N application)	Increase / decrease of yield over T_5 (%)				Total Soluble Solids (^{0}Brix)			
	2002	2003	2004	Average	2002	2003	2004	Average
T_1(Entire dose in July)	+1.22	+ 7.36	+1.63	+3.4	15.95	17.72	19.87	17.85
T_2 (Two splits : July & March)	+14.85	+ 26.89	+11.03	+17.6	16.10	18.07	21.56	18.58
T_3 (Two splits : July & Sept.)	-1.56	- 3.72	-2.61	-2.6	15.55	17.47	20.33	17.78
T_4(Three splits : June, Sept. & March)	+4.99	+ 5.35	+3.56	+4.6	16.60	17.97	21.07	18.55
T_5(Control : June & August, Normal practice in WB)	-	-	-	-	16.05	17.77	20.42	18.08
SEm($\pm$)	-	-	-	-	0.591	0.123	0.334	-
C.D. at 5%	-	-	-	-	N.S.	0.378	1.028	-

NS= Non-significant

Table 4.1.16 : Quality of fruit as influenced by different timing of nitrogen application.

Treatment (Time of N application)	Reducing sugar (% fresh wt.)				Total sugar (% fresh wt.)			
	2002	2003	2004	Average	2002	2003	2004	Average
T_1(Entire dose in July)	14.20	13.27	13.15	13.54	15.29	15.37	15.50	15.39
T_2 (Two splits : July & March)	14.25	13.89	13.91	14.02	15.42	15.86	15.94	15.74
T_3 (Two splits : July & Sept.)	13.80	13.87	13.53	13.73	14.97	15.31	15.35	15.21
T_4(Three splits : June, Sept. & March)	14.42	13.82	13.73	13.99	15.81	15.72	15.80	15.71
T_5(Control : June & August, Normal practice in WB)	14.19	13.88	13.61	13.89	15.34	14.99	15.64	15.32
SEm(±)	0.402	0.397	1.606	-	0.544	0.132	0.127	-
C.D. at 5%	N.S.	NS	NS	-	N.S.	0.408	0.391	-

NS= Non-significant

Table 4.1.17 :Ascorbic acid (Vitamin C) content and TSS/acid ratio of fruit.

Treatment (Time of N application)	Ascorbic acid (mg/100g pulp)				TSS/acid ratio			
	2002	2003	2004	Average	2002	2003	2004	Average
T_1(Entire dose in July)	16.46	19.86	24.27	20.20	34.31	32.22	36.60	34.38
T_2 (Two splits : July & March)	9.16	27.42	28.62	32.60	30.93	42.02	44.94	39.30
T_3 (Two splits : July & Sept.)	17.46	20.34	25.35	21.05	27.35	36.40	39.75	34.50
T_4(Three splits : June, Sept. & March)	20.75	25.19	24.91	23.62	33.59	39.07	42.66	38.44
T_5(Control : June & August, Normal practice in WB)	10.36	21.44	23.97	18.59	41.21	34.84	39.48	38.18
SEm(±)	5.823	11.45	8.382	-	-	-	-	-
C.D. at 5%	N.S.	NS	NS	-	-	-	-	-

NS= Non-significant

Table 4.1.18 : Correlation matrix of relationship between pre-flowering leaf N content, shoot growth rate, carbohydrate content of leaf and shoot, flowering and yield.

Variables	Leaf N (Oct.-Dec.)	Shoot growth rate (Oct.-Dec.)	Leaf carbohyd -rate (Dec.)	Shoot carbohyd -rate (Dec.)	Flower -ing shoot (%)	Yield (kg/ha)
Leaf N (Oct.-Dec.)	1.000					
Shoot growth rate (Oct.-Dec.)	0.979 **	1.000				
Leaf carbohydrate (Dec.)	-0.955 *	-0.995 **	1.000			
Shoot carbohydrate (Dec.)	-0.920 *	-0.964 **	0.977 **	1.000		
Flowering shoot (%)	-0.972 **	-0.997 **	0.993 **	0.975 **	1.000	
Yield (kg/ha)	-0.888 *	-0.943 *	0.996 **	0.996 **	0.961 **	1.000

** Significant at P= 0.01,
* Significant at P= 0.05

Table 4.1.19 : Correlation matrix of relationship between post-flowering leaf N content, carbohydrate content of leaf and shoot, fruit retention, yield and quality (TSS/acid ratio) of fruit.

Variables	Leaf N (Apr.-May)	Leaf carbohyd-rate (Mar.)	Shoot carbohyd-rate (Mar.)	Fruit retent-ion (%)	Yield (kg/ha)	Quality (TSS/acid ratio)
Leaf N (Apr.-May)	1.000					
Leaf carbohydrate (Mar.)	0.425	1.000				
Shoot carbohydrate (Mar.)	0.884 *	0.712	1.000			
Fruit retention (%)	0.968 **	0.257	0.792	1.000		
Yield (kg/ha)	0.841	0.712	0.990 **	0.725	1.000	
Quality (TSS/acid ratio)	0.463	0.826	0.583	0.249	0.604	1.000

** Significant at P= 0.01,
* Significant at P= 0.05

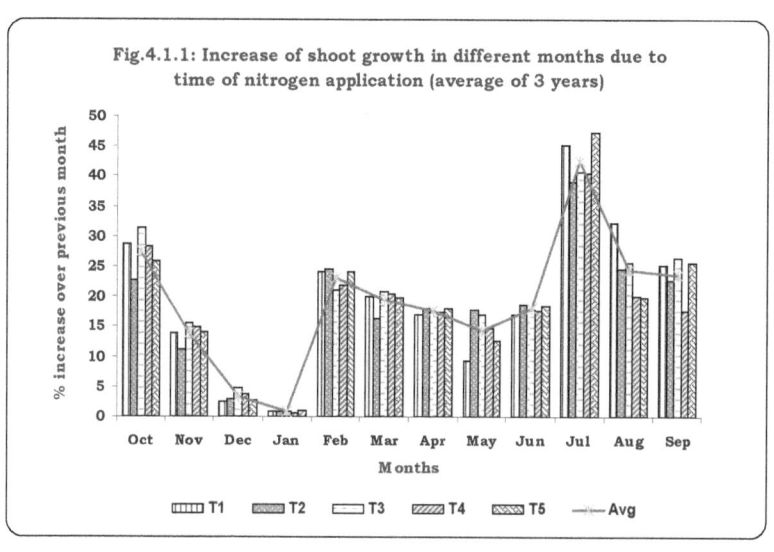

Fig.4.1.1: Increase of shoot growth in different months due to time of nitrogen application (average of 3 years)

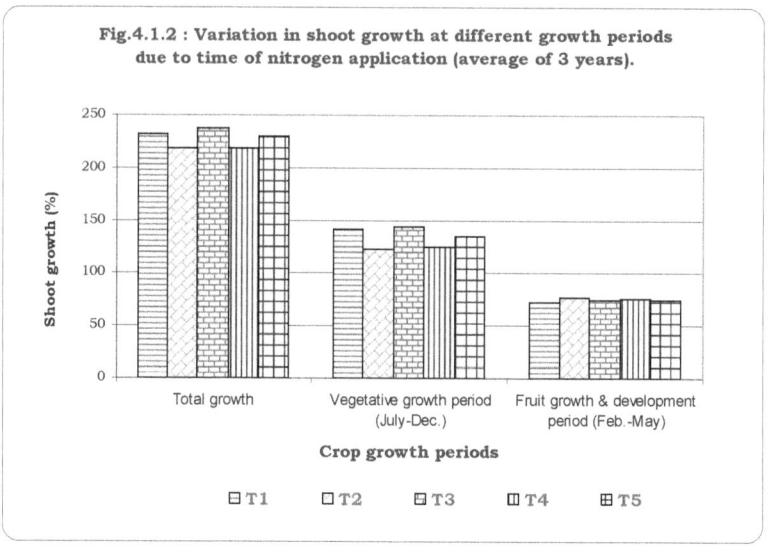

Fig.4.1.2 : Variation in shoot growth at different growth periods due to time of nitrogen application (average of 3 years).

81

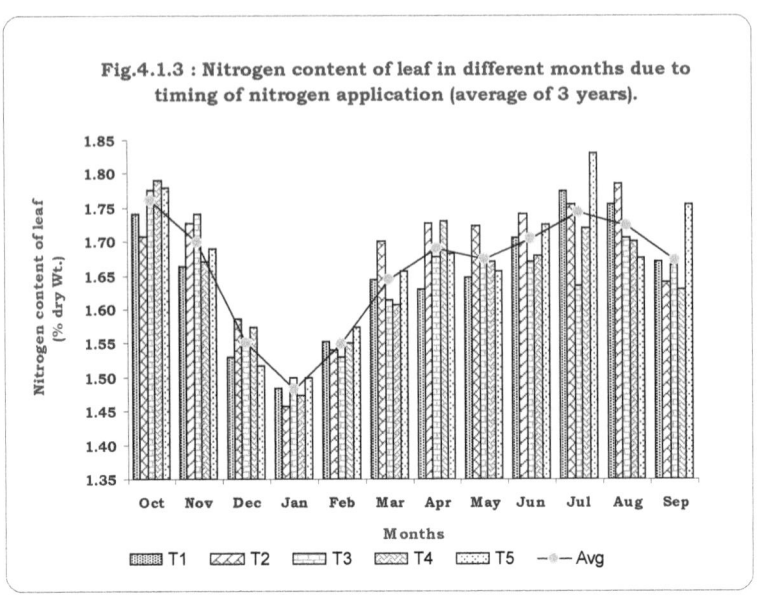

Fig.4.1.3 : Nitrogen content of leaf in different months due to timing of nitrogen application (average of 3 years).

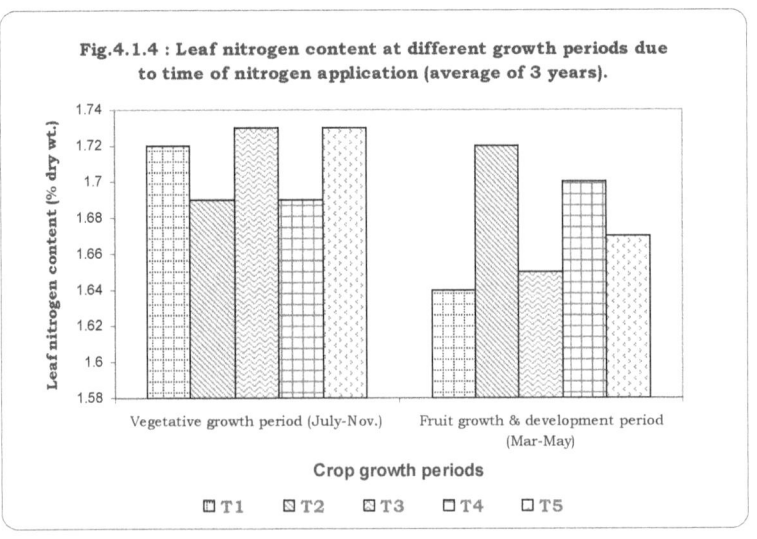

Fig.4.1.4 : Leaf nitrogen content at different growth periods due to time of nitrogen application (average of 3 years).

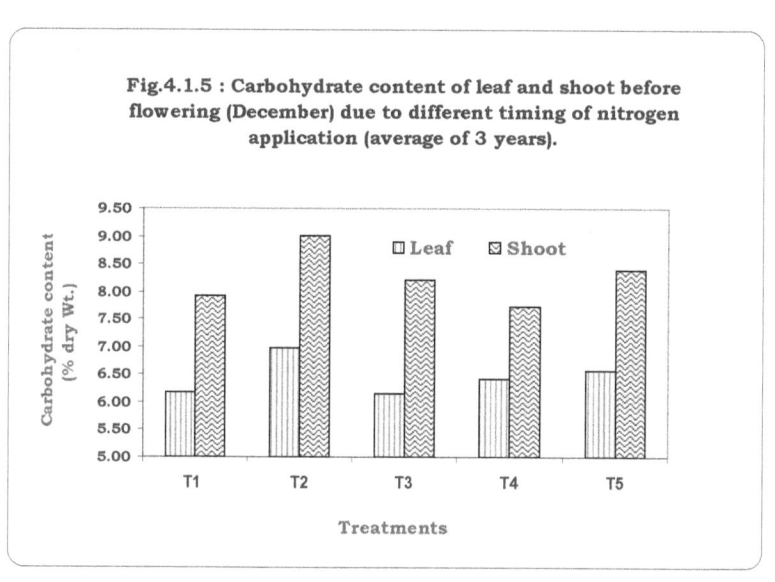

Fig.4.1.5 : Carbohydrate content of leaf and shoot before flowering (December) due to different timing of nitrogen application (average of 3 years).

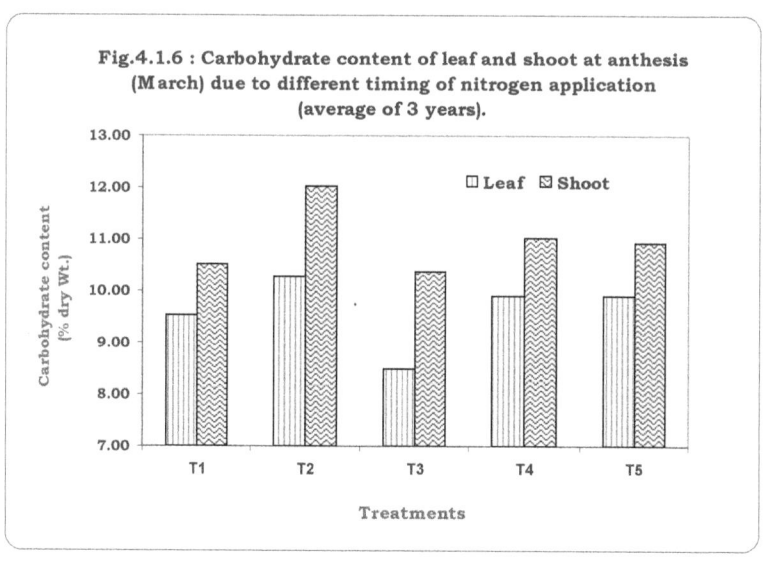

Fig.4.1.6 : Carbohydrate content of leaf and shoot at anthesis (March) due to different timing of nitrogen application (average of 3 years).

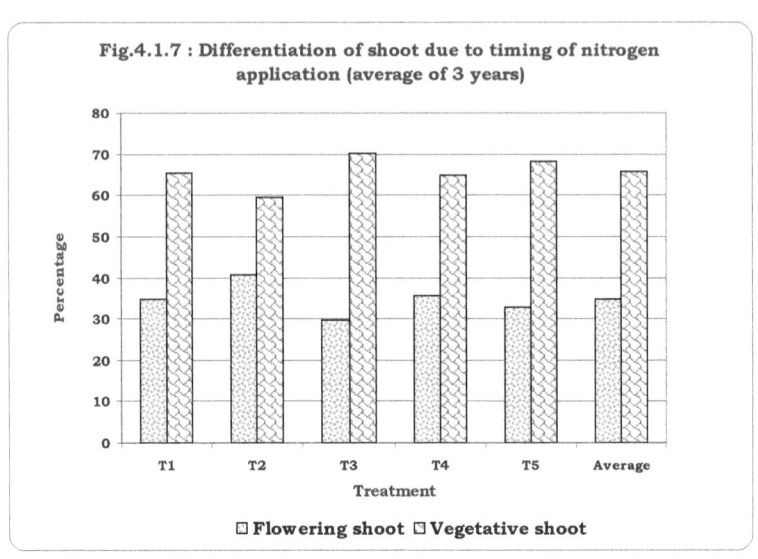

Fig.4.1.7 : Differentiation of shoot due to timing of nitrogen application (average of 3 years)

□ Flowering shoot □ Vegetative shoot

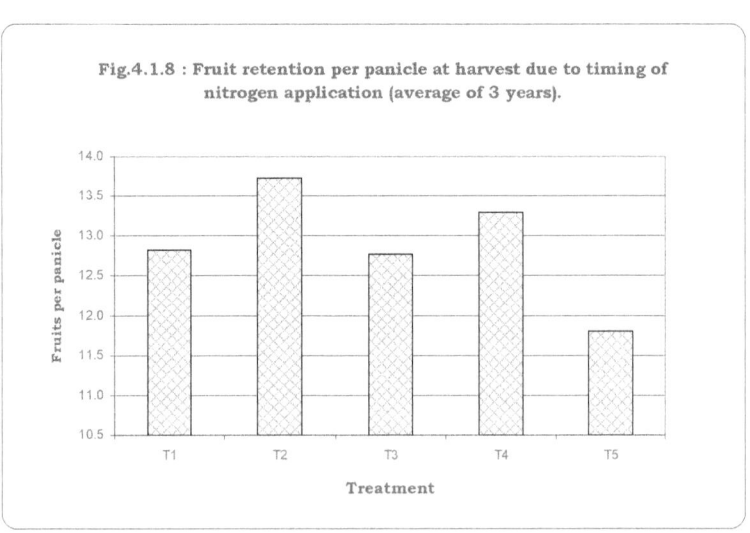

Fig.4.1.8 : Fruit retention per panicle at harvest due to timing of nitrogen application (average of 3 years).

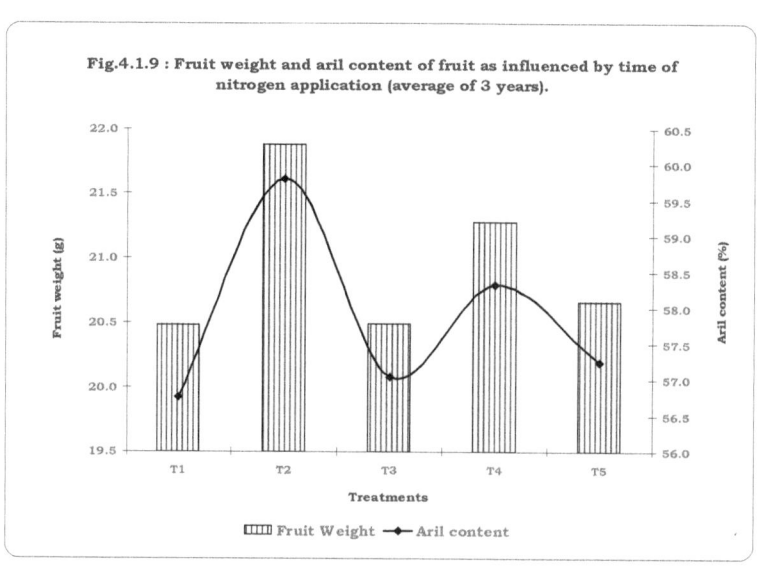

Fig.4.1.9 : Fruit weight and aril content of fruit as influenced by time of nitrogen application (average of 3 years).

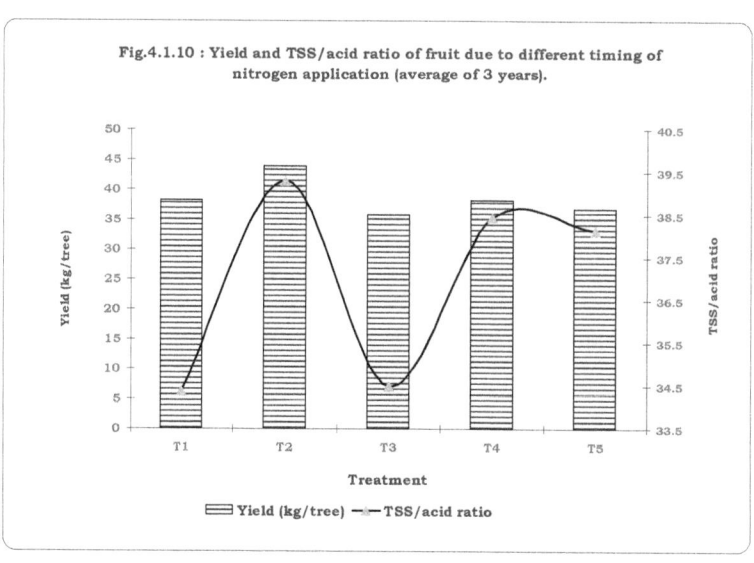

Fig.4.1.10 : Yield and TSS/acid ratio of fruit due to different timing of nitrogen application (average of 3 years).

Experiment II : Refinement of leaf and soil sampling

Formulation of a fertilizer programme on the basis of leaf and soil analysis is advantageous and therefore, becoming popular in fruit cultivation. In this approach, however, it is essential to standardize the index tissue, sampling techniques of leaf (index tissue) and soil and the adequate leaf nutrients ranges or standards for effective recommendation to be made. Recommendations are expected to be extremely precise and effective, only if the factors of sampling error are minimized to the least possible levels. The horrible consequence of wrong interpretation may result due to sampling error and faulty diagnosis. In this experiment, different variables of leaf (direction, age/position on the shoot, canopy height) and soil (depth and crop growth periods) sampling were studied.

4.2.1 Study of leaf variables

Direction: The leaf nutrient content in a tree showed significant variations in different months and due to different directions of sampling (Table 4.2.1). In general, the leaf nutrients (N, P and K) content were lower during dormancy (January) compared with fruit growth and development period (March to May) and post harvest vegetative growth period (July to November). The leaf sampled from east direction of the tree showed the highest content of N, P and K as compared with west, south and north directions, irrespective of time of sampling (Fig. 4.2.1).

During fruiting period (March-May), the N, P and K content of leaf in east direction were estimated as 1.69%, 0.18% and 1.10%, respectively, which was found lower (1.54%, 0.15% and 0.98%, respectively) in the north direction. At that period, the leaf nutrients (N, P and K) content in west and south directions varied between 1.60% and 1.65%, 0.16% and .18% and 1.00% and 1.10%, respectively.

Between July and November (vegetative growth period), the N, P and K content of leaf showed almost similar trends of variation due to directions and varied between 1.48% and 1.71%, 0.13% and 0.20% and 0.87% and 1.18%, respectively. The leaf N content recorded between 1.60% and 1.71%

in east direction compared with 1.48% to 1.60% in north direction. In west and south directions, the leaf N content varied between 1.50% and 1.70%. The variations in P and K content of leaf in all the four directions were found lower between July and September compared with the leaf P and K content between September and November. The highest content of leaf P (0.19% to 0.20%) and K (1.10% to 1.18%) were found in east direction compared with 0.16% to 0.18% and 1.00% to 1.01%, respectively in north direction.

Leaf nutrients (N, P and K) content, in general, declined from November and reached minimum in January, irrespective of the direction of sampling. In east direction, the N, P and K content of leaf in the month of January were 1.40%, 0.14% and 0.78%, respectively compared with 1.32%, 0.11% and 0.68%, respectively in the north direction. For west and south directions, it ranged between 1.36% and 1.40%, 0.12% and 0.13% and 0.71% to 0.76%, respectively.

Position of leaf on shoot (leaf age) : The N, P and K content of leaf showed significant variations due to the position of leaf on the shoot or age of leaf (Table 4.2.1). In general, the position of leaf showed higher variations in N (1.50%-1.72%), P (0.14%-0.20%) and K (0.95%-1.18%) content during fruit development period (March-May) and post harvest vegetative growth period i.e., between July and November (1.41%-1.70%, 0.15%-0.19% and 0.75%-1.19%, respectively) compared with dormancy period in January (1.36%-1.41%, 0.12%-0.14% and 0.73%-0.75%, respectively) (Fig. 4.2.2).

During fruit development period (March to May), the second leaf showed maximum content of leaf N (1.70%-1.72%), P (0.20%-0.21%) and K (1.10%-1.18%) which were minimum (1.50%-1.65%, 0.14%-0.18% and 0.95%-0.98%, respectively) in the sixth leaf. However, the first, third and fourth leaves showed very little variations in N, P and K content in the month of March (1.70%-1.71%, 0.19%-0.20% and 1.11%-1.15%, respectively) and May (1.60%-1.68%, 0.16%-0.18% and 0.98%-1.00%, respectively).

During peak vegetative growth period (July to September), the leaf N, P and K content varied between 1.50% and 1.70%, 0.16% and 0.19% and 0.81% and 1.19%, respectively due to the position of leaf. From the second leaf to sixth leaf, there were gradual decline in N (1.70% to 1.50%), P (0.19% to 0.16%) and K (1.19% to 0.81%) contents, while the variations between first and third leaves were not very pronounced in the month of July (1.67%-1.68%, 0.18% and 1.17%-1.19%, respectively) and September (1.68%-1.70%, 0.18% and 0.99%-0.98%, respectively).

Height of sampling :Leaves from different height on the tree canopy significantly influenced the nutrient composition (Table 4.2.1). In general, the nutrients (N, P and K) content of leaf within 1 m of canopy height (from ground level) were low (1.30%-1.61%, 0.12%-0.16% and 0.72%-1.10%, respectively) which gradually increased with sampling height up to 3 m (1.38%-1.71%, 0.15%-0.20% and 0.77%-1.18%, respectively) and then declined marginally up to the sampling height of 5 m or above (1.51%-1.68%, 0.14%-0.19% and 0.73%-1.14%, respectively) (Fig. 4.2.3).

Leaves when estimated at a height of 2 to 5 m during fruit growth period (March) showed higher content of leaf N (1.68%-1.71%), P (0.18%-0.20%) and K (1.13%-1.17%) content compared with 1 m and more than 5 m sampling height (1.61%-1.65%, 0.15%-0.18% and 1.10%-1.13%, respectively). It appears that canopy height of 2 to 5 m supposed to better development of fruit.

Between July and September (vegetative growth period), maximum leaf N (1.69%-1.71%), P (0.20%) and K (1.15%-1.18%) contents were recorded at a height of 3 m. However, the variations in leaf nutrient (N, P and K) contents at 2, 4 and 5 m were not very pronounced compared with the leaves near ground level (1.53%-1.60%, 0.16% and 1.10%, respectively) and above 5 m (1.61%-1.67%, 0.17%-0.18 and 1.12%-1.13%, respectively).

4.2.2 Study of soil nutrient content at different soil depths during growth and development periods

Organic carbon : In general, the organic carbon content of 0-90 cm soil profile increased gradually from 0.536% in January to 0.632% in July and then declined in the following months (Table 4.2.2, Fig. 4.2.4). The average of soil organic carbon content in different growth and development periods showed that the organic carbon content was highest (0.621%) in the top 30 cm soil profile which declined to 0.605% and 0.583% at 60 and 90 cm soil depths, respectively.

During fruit development period (March to May), the soil organic carbon content varied between 0.591% and 0.640%, 0.593% and 0.623% and 0.573% and 0.612% at 30, 60 and 90 cm soil depths, respectively. The organic carbon content was comparatively higher at the vegetative growth period (July to November) which varied between 0.633% and 0.650%, 0.624 and 0.635% and 0.611% and 0.582% at 30, 60 and 90 cm soil depths, respectively.

Available nitrogen :The available nitrogen content of 0-90 cm soil profile was recorded 312.3 kg/ha at dormancy (January) which declined during fruit development period (March-May) to 257.1 kg/ha and then increased in the following months (Table 4.2.2, Fig. 4.2.4). The upper soil profile (0-60 cm) was found rich in available nitrogen content (283.5-285.8 kg/ha) compared with 278.9 kg/ha at 90 cm soil depth.

Between January and May (flowering and fruit development period), the available soil nitrogen status showed the maximum depletion from 314.2 to 261.4kg, 313.2 to 258.6kg and 309.3 to 251.4 kg/ha at 30, 60 and 90 cm soil depths, respectively indicating that fruiting was a nutrient depleting process. The available nitrogen content increased in the month of July (273.2, 265.3 and 262.6 kg/ha at 30, 60 and 90 cm depths, respectively) following fertilization after fruit harvest, which declined to 263.3, 268.4 and 261.2 kg/ha at 30, 60 and 90 cm depths, respectively in September and then increased again between November and January (295.4 and 314.2, 294.2 and 313.2 and 290.6 and 309.3 kg/ha, respectively).

Available phosphorus : Sampling at different growth and development periods showed considerable variations in available phosphorus content of soil ranging between 41.57 kg/ha in May and 63.85 kg/ha in January (Table 4.2.2, Fig. 4.2.4). Average available soil phosphorus content in different months was recorded maximum (54.00 kg/ha) at 30 cm soil profile which declined to 52.96 and 50.89 kg/ha at 60 and 90 cm soil depths, respectively.

The soil sampled in January at 30, 60 and 90 cm depths showed maximum (65.62, 64.25 and 61.68 kg/ha, respectively) content of available phosphorus which sharply declined during flowering and fruit development period to 42.53, 41.63 and 40.56 kg/ha at 30, 60 and 90 cm depths, respectively in May and then increased in July (55.33, 53.45 and 51.69 kg/ha, respectively) after fertilization. In September, the available soil phosphorus content was comparatively low at different depths (43.65, 43.57 and 40.24 kg/ha, respectively) compared with that recorded in the month of November (54.15, 52.35 and 50.57 kg/ha, respectively).

Available potassium : Variations in available potassium content of soil showed almost similar trends with that of available soil nitrogen content due to depth and time of sampling. The available potassium content of 0-90 cm soil profile was recorded 143.6 kg/ha at dormancy (January) which declined during fruiting period (March-May) to 132.4 kg/ha and then increased in July (135.4 kg/ha) following fertilization after fruit harvest (Table 4.2.2, Fig. 4.2.4). The upper soil profile (0-60 cm) was found rich in available potassium content (136.0-136.9 kg/ha) compared with 132.8 kg/ha at 90 cm soil profile.

During panicle formation to fruit harvest, the available potassium content of soil declined from 146.6, 143.8 and 140.4 kg/ha at 30, 60 and 90 cm soil depths, respectively in January to 124.1, 123.6 and 120.3 kg/ha, respectively in the month of May. Fertilization after fruit harvest increased the available potassium content of soil in July (137.2, 135.5 and 133.4 kg/ha at 30, 60 and 90 cm depths, respectively), which slightly declined in the month of September (134.6, 133.2 and 129.2 kg/ha, respectively) and then again increased in the following months until January.

4.2.3 Adequate leaf nutrient range (leaf nutrient standards) :

The flowering and yield under different treatments of Experiment-I were studied. The results of Experiment-I revealed that the leaf nutrient content, especially the nitrogen and carbohydrate content of leaf and shoot had pronounced effect on flowering and yield. The T_2 treatment (total N, P_2O_5 and K_2O at two equal splits i.e., in July and March) showed maximum flowering shoots (40.60%), highest yield (43.83 kg/tree) and better quality (TSS/acid ratio 39.30) of fruit over other treatments. The trees under T_2 treatment were studied to establish the adequate leaf nutrients range.

At pre-flowering stage (December), a composite sample of 16-20 pair of leaflets were collected from four directions (east, west, north and south) at the canopy height between 2 m and 4 m from ground, taking the second pair of leaflets of the second leaf from tip of the shoot. The mature leaf (2nd pair) just behind the panicle at anthesis was sampled similarly as the suitable 'index tissue' for estimating nutrition status of bearing litchi tree cv. Bombai.

Based on flowering and yield performance, the pre-flowering (December) optimum leaf nutrient levels and C/N ratio suitable for flower bud differentiation were established as :

N	: 1.40 - 1.50 %	Fe	: 50 – 85 ppm
P	: 0.14 - 0.17 %	Mn	: 180 - 200 ppm
K	: 0.80 - 1.00 %	Zn	: 18 – 24 ppm
Ca	: 0.90 – 1.00 %	Cu	: 8 – 12 ppm
Mg	: 0.40 – 0.50 %	B	: 22 - 28 ppm
S	: 0.18 - 0.24 %	Na	: < 300 ppm
Leaf carbohydrate : 6.5-7.5 %		C/N ratio : 4.5-5.5	
Shoot carbohydrate : 9-10 %		C/N ratio : 7.5-8.5	

The adequate leaf nutrient levels and C/N ratio at anthesis for optimum yield and better quality of fruit were found as :

N	: 1.65 - 1.75 %	Fe	: 80 - 100 ppm
P	: 0.18 - 0.20 %	Mn	: 180 - 200 ppm
K	: 1.00 - 1.20 %	Zn	: 15 - 25 ppm

Ca	: 0.80 - 1.00 %	Cu	: 10 - 15 ppm
Mg	: 0.40 - 0.50 %	B	: 25 - 40 ppm
S	: 0.20 - 0.25 %	Na	: < 300 ppm
		C/N ratio : 6-7	

Leaf carbohydrate : 9.5-10.5 %

Shoot carbohydrate : 12-13 %   C/N ratio : 9.5-10.5

4.2.4 DISCUSSION

Considerable variations in leaf nutrients composition and soil nutrients content were recorded due to the variables of leaf and soil sampling. The leaf nutrient (N, P and K) content, in general, were lower during vegetative dormancy (January) compared with fruit growth and development period (March to May) and post harvest vegetative growth period (July to November). The effect of climatic variations on certain physiological processes of perennial plant is supposed to influence the nutritional composition of a plant in different seasons (Martin Prevel et al., 1984). This trend of seasonal variations in leaf N, P and K content were observed irrespective of sampling direction, position of leaf on shoot (leaf age) and height of sampling.

During fruiting period (March-May), the N, P and K content of leaf in east direction were estimated as 1.69%, 0.18% and 1.10%, respectively, which was found lower (1.54%, 0.15% and 0.98%, respectively) in the north direction. In this period, the leaf nutrients (N, P and K) content in west and south directions varied between 1.60% and 1.65%, 0.16% and 0.18% and 1.00% and 1.10%, respectively. Sanyal and Mitra (1990) also reported that the N, P and K content of leaf was higher in the east direction compared with west, north and south directions and the influence of direction on nutrient accumulation in leaf was attributed to the relative exposure of leaf to light and simultaneously its effect on the photosynthetic process. A leaf receiving direct sunlight was found nine to ten times more productive than one in diffused light (Heinicke, 1976).

In general, the position of leaf showed higher variations in N (1.50%-1.72%), P (0.14%-0.20%) and K (0.95%-1.18%) content during fruit development period (March-May) and post harvest vegetative growth period i.e., between July and November (1.41%-1.70%, 0.15%-0.19% and 0.75%-1.19%, respectively) compared with dormancy period in January (1.36%-1.41%, 0.12%-0.14% and 0.73%-0.75%, respectively). During fruit development period (March to May), the second leaf showed maximum content of leaf N (1.70%-1.72%), P (0.20%-0.21%) and K (1.10%-1.18%) which were minimum (1.50%-1.65%, 0.14%-0.18% and 0.95%-0.98%, respectively) in the sixth leaf. However, the first, third and fourth leaves showed very little variations in N, P and K content in the month of March (1.70%-1.71%, 0.19%-0.20% and 1.11%-1.15%, respectively) and May (1.60%-1.68%, 0.16%-0.18% and 0.98%-1.00%, respectively). Similar trend of variation in leaf N, P and K content due to leaf age was observed during vegetative growth period and dormancy (January). Sanyal and Mitra (1990) also reported that the N, P and K concentration in leaf decreased as the leaves aged while Ca, Mg and S were found to accumulate on mature leaves. According to Robson (1981), the N, P and K being highly mobile in plants, their concentrations decreased with leaf age. Many scientists therefore, preferred the recently mature leaf for sampling of N, P and K estimation in litchi (Menzel et al., 1987, Sanyal and Mitra, 1990).

The nutrient (N, P and K) content of leaf within 1 m of canopy height were low (1.30%-1.61%, 0.12%-0.16% and 0.72%-1.10%, respectively) which gradually increased with sampling height up to 3 m (1.38%-1.71%, 0.15%-0.20% and 0.77%-1.18%, respectively) and then declined marginally up to the sampling height of 5 m or above (1.51%-1.68%, 0.14%-0.19% and 0.73%-1.14%, respectively). Leaves when estimated at a height of 2 to 5 m during fruit growth period (March) showed higher content of leaf N (1.68%-1.71%), P (0.18%-0.20%) and K (1.13%-1.17%) compared with 1 m and more than 5 m sampling height (1.61%-1.65%, 0.15%-0.18% and 1.10%-1.13%, respectively). Sayal et al. (1999) observed that the litchi fruits at lower half portion of the tree were heavier in weight compared with those collected from upper half portion of tree. It appears that canopy height of 2 to 5 m (i.e., higher leaf N, P and K content) supposed to better development of fruit and yield (Hasan and Chattapadhyay, 1997) which could be attributed to the spatial orientation of shoots and relative exposure to sunlight. However, low N, P and K content of leaf on lopping branches within 1 m of canopy height might be due to relatively less exposure to sunlight and photosynthetic activity (Heincke, 1976).

The organic carbon, available N, P and K in soil, in general, declined gradually from 0.621%, 285.8 kg/ha, 54.00 kg/ha and 136.9 kg/ha, respectively at 30 cm depth to 0.583%, 278.9 kg N/ha, 50.89 kg P/ha and 132.8 kg K/ha, respectively at 90 cm soil depth. Gogoi et al. (2003) also recorded similar variation in organic carbon, available N, P and K content of soil due to depth in subtropical Brahmaputra plain of Assam and reported maximum organic carbon (1.65%), available N (302.8 kg/ha), P (76.73 kg/ha) and K (145.13 kg/ha) at 0-20 cm soil depth compared with 0.58% organic carbon, 247.6 kg N/ha, 39.40 kg P/ha and 30.30 kg K/ha at 40-60 cm soil depth. In the present investigation, the top (0-30 cm) soil layer showed highest concentration of organic carbon (0.621%), available N (285.8 kg/ha), P (54.00 kg/ha) and K (136.9 kg/ha) which is supposed due to decomposition of weeds and fallen leaves and application and incorporation of fertilizers in the top soil.

Notable seasonal variation was recorded in organic carbon, available N, P and K content of soil. The organic carbon content was minimum (0.536%) in winter (January) compared with 0.632% to 0.627% in monsoon (July to September). This seasonal variation in soil organic carbon content might be related with the decomposition rate of organic debris which is favoured by monsoon (high temperature and moisture) and slowed down in winter (low temperature and humidity). The available N, P and K content of soil showed almost similar trend of seasonal variation which were recorded maximum (312.3 kg/ha, 63.85 kg/ha and 143.6 kg/ha, respectively) during vegetative dormancy (January). During fruit growth and development (March to May), the available N, P and K content of soil showed gradual decline and then increased in July due to fertilization after fruit harvest. The concentration of available N, P and K in soil decreased again during post harvest vegetative growth period (July to September). These two periods of nutrients (available

N, P and K) depletion might be related with the maximum uptake of these nutrients by vigorously growing post harvest vegetative growth and growth and development of fruits (Gogoi et al., 2003), while slow rate of uptake during vegetative dormancy (winter) could have resulted in increased concentration of these nutrients in soil.

Based on the correlation studies on pre-flowering nutrient status with flowering and post-flowering nutrient status with yield and quality (Experiment-I), the adequate leaf nutrient levels and C/N ratio in leaf and shoot were estimated. A high C/N ratio in leaf and shoot before flowering favoured flowering in litchi (Koen and Smart, 1982; Menzel and Simpson 1990; Chen et al., 1994 and Davenport et al., 1999). The mature leaf adjacent to the panicle at anthesis was found suitable as the index tissue. Koen and Smart (1983) from South Africa also suggested to sample the second leaf below the developing fruit cluster. Our estimates of optimum leaf nutrient ranges showed close approximation with the recommended standards suggested by Kadman and Slor (1982) from Israel, Menzel et al. (1992c) from Australia, Koen et al. (1981) from South Africa and Huang et al. (1998) from Taiwan. Some comparable variations, however, do exist probably due to the variability in soil and climatic conditions of different places and the variety under study.

Plate 5: PHOTOGRAPHS SHOWING LEAF SAMPLING TECHNIQUE IN LITCHI CV. BOMBAI

Table 4.2.1(a) : Influence of leaf variables on nutrient content of leaf (% dry weight).

Leaf variables		September			November			January		
		N %	P %	K %	N %	P %	K %	N %	P %	K %
Direction	East	1.70	0.20	1.10	1.60	0.16	0.96	1.40	0.14	0.78
	West	1.65	0.18	1.08	1.50	0.14	0.89	1.36	0.12	0.71
	North	1.60	0.18	1.01	1.48	0.13	0.87	1.32	0.11	0.68
	South	1.65	0.20	1.10	1.52	0.15	0.90	1.40	0.13	0.76
	SEm (±)	0.018	0.008	0.009	0.043	0.004	0.030	0.073	0.011	0.015
	C.D. at 5%	0.041	0.019	0.021	0.093	0.010	0.067	0.081	0.026	0.035
Age (Position of leaf on the shoot)	1st leaf	1.70	0.18	0.98	1.56	0.16	0.86	1.41	0.14	0.75
	2nd leaf	1.70	0.19	1.10	1.58	0.17	0.90	1.41	0.14	0.76
	3rd leaf	1.68	0.18	0.99	1.56	0.17	0.88	1.41	0.13	0.75
	4th leaf	1.60	0.17	0.95	1.48	0.16	0.87	1.38	0.13	0.74
	5th leaf	1.58	0.17	0.84	1.42	0.15	0.81	1.36	0.12	0.73
	6th leaf	1.50	0.16	0.81	1.41	0.15	0.75	1.36	0.12	0.73
	SEm (±)	0.040	0.018	0.024	0.091	0.012	0.058	0.076	0.025	0.031
	C.D. at 5%	0.018	0.008	0.011	0.040	0.005	0.025	0.034	0.011	0.013
Height of sampling (from ground level)	1 m	1.60	0.16	1.10	1.50	0.11	0.90	1.30	0.12	0.72
	2 m	1.63	0.18	1.14	1.55	0.13	0.94	1.35	0.13	0.75
	3 m	1.69	0.20	1.15	1.57	0.15	0.95	1.38	0.15	0.77
	4 m	1.65	0.19	1.15	1.54	0.14	0.95	1.37	0.15	0.75
	5 m	1.65	0.18	1.13	1.55	0.14	0.93	1.36	0.15	0.74
	>5 m	1.61	0.18	1.12	1.51	0.14	0.92	1.36	0.14	0.73
	SEm (±)	0.031	0.020	0.021	0.097	0.013	0.065	0.073	0.019	0.035
	C.D. at 5%	0.014	0.009	0.009	0.043	0.006	0.029	0.032	0.008	0.015

Table 4.2.1(b) : Influence of leaf variables on nutrient content of leaf (% dry weight).

Leaf variables		March			May			July		
		N %	P %	K %	N %	P %	K %	N %	P %	K %
Direction	East	1.70	0.19	1.11	1.68	0.18	1.10	1.71	0.19	1.18
	West	1.60	0.17	1.00	1.60	0.17	1.10	1.70	0.18	1.11
	North	1.58	0.16	0.95	1.50	0.15	1.02	1.60	0.16	1.00
	South	1.65	0.18	1.10	1.60	0.16	1.06	1.65	0.17	1.10
	SEm (±)	0.036	0.014	0.045	0.029	0.008	0.037	0.041	0.012	0.024
	C.D. at 5%	0.081	0.031	0.101	0.066	0.020	0.083	0.092	0.028	0.055
Age (Position of leaf on the shoot)	1st leaf	1.70	0.20	1.15	1.65	0.18	1.00	1.68	0.18	1.17
	2nd leaf	1.72	0.21	1.18	1.70	0.20	1.10	1.70	0.19	1.19
	3rd leaf	1.71	0.20	1.15	1.68	0.18	1.00	1.67	0.18	1.19
	4th leaf	1.70	0.19	1.11	1.60	0.16	0.98	1.65	0.18	1.17
	5th leaf	1.68	0.19	1.00	1.56	0.15	0.96	1.63	0.17	1.16
	6th leaf	1.65	0.18	0.98	1.50	0.14	0.95	1.62	0.16	1.15
	SEm (±)	0.076	0.023	0.096	0.055	0.019	0.085	0.011	0.029	0.061
	C.D. at 5%	0.034	0.010	0.042	0.024	0.008	0.038	0.005	0.013	0.027
Height of sampling (from ground level)	1 m	1.61	0.15	1.10	1.60	0.13	0.90	1.53	0.16	1.10
	2 m	1.69	0.18	1.16	1.65	0.16	1.05	1.69	0.19	1.15
	3 m	1.71	0.20	1.17	1.69	0.17	1.08	1.71	0.20	1.18
	4 m	1.70	0.19	1.15	1.63	0.16	1.01	1.70	0.18	1.16
	5 m	1.68	0.19	1.13	1.62	0.15	0.96	1.68	0.17	1.14
	>5 m	1.65	0.18	1.13	1.60	0.15	0.94	1.67	0.17	1.13
	SEm (±)	0.067	0.022	0.084	0.065	0.024	0.071	0.015	0.030	0.076
	C.D. at 5%	0.030	0.010	0.037	0.029	0.011	0.031	0.00	0.013	0.034

Table 4.2.2 (a): Soil nutrient content at different soil depths during growth and development periods.

Month	Organic Carbon (%)				Available Nitrogen (kg/ha)			
	30 cm	60 cm	90 cm	Average	30 cm	60 cm	90 cm	Average
September	0.642	0.632	0.606	0.627	263.3	268.4	261.2	264.3
November	0.633	0.624	0.582	0.613	295.4	294.2	290.6	293.4
January	0.567	0.525	0.516	0.536	314.2	313.2	309.3	312.3
March	0.591	0.593	0.573	0.586	307.3	301.4	298.5	302.4
May	0.640	0.623	0.612	0.625	261.4	258.6	251.4	257.1
July	0.650	0.635	0.611	0.632	273.2	265.3	262.6	267.0
Average	0.621	0.605	0.583	-	285.8	283.5	278.9	-

Table 4.2.2 (b): Soil nutrient content at different soil depths during growth and development periods.

Month	Available Phosphorus (kg/ha)				Available Potassium (kg/ha)			
	30 cm	60 cm	90 cm	Average	30 cm	60 cm	90 cm	Average
September	43.65	43.57	40.24	42.49	134.6	133.2	129.2	132.3
November	54.15	52.35	50.57	52.36	137.5	137.3	135.3	136.7
January	65.62	64.26	61.68	63.85	146.6	143.8	140.4	143.6
March	62.53	62.51	60.59	61.88	141.5	140.7	138.2	140.1
May	42.53	41.63	40.56	41.57	124.1	125.6	120.3	132.4
July	55.53	53.45	51.69	53.56	137.2	135.5	133.4	135.4
Average	54.00	52.96	50.89	-	136.9	136.0	132.8	-

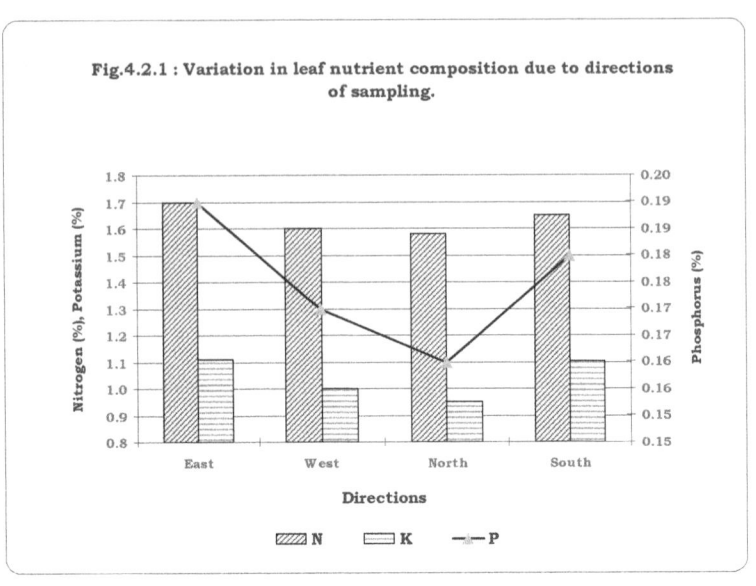

Fig.4.2.1 : Variation in leaf nutrient composition due to directions of sampling.

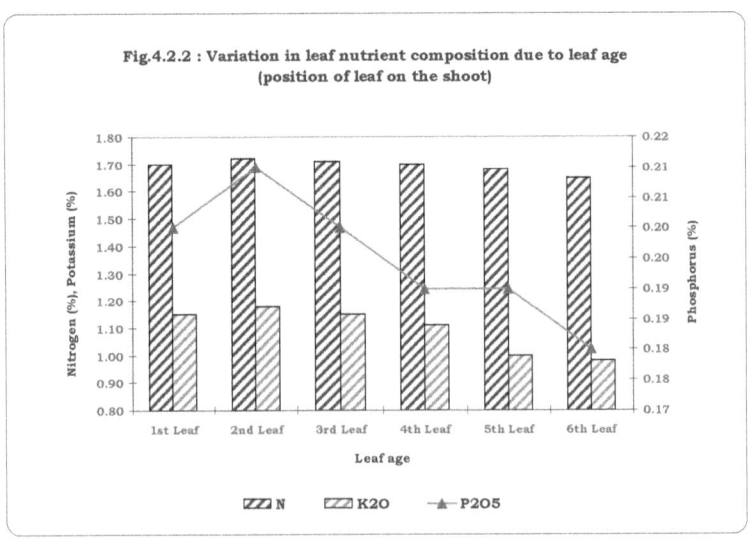

Fig.4.2.2 : Variation in leaf nutrient composition due to leaf age (position of leaf on the shoot)

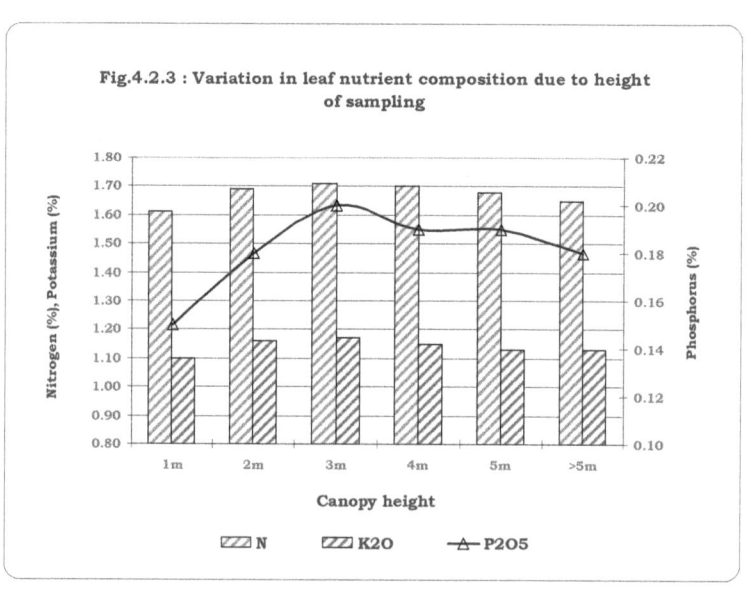

Fig.4.2.3 : Variation in leaf nutrient composition due to height of sampling

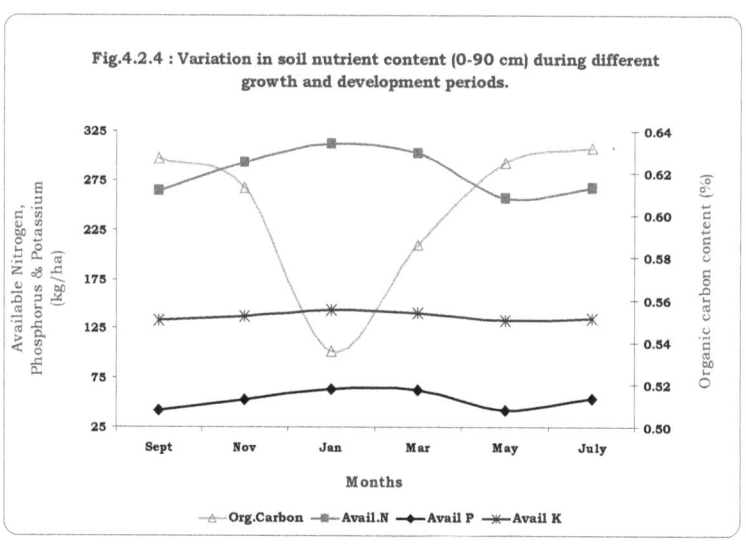

Fig.4.2.4 : Variation in soil nutrient content (0-90 cm) during different growth and development periods.

Experiment III : To study the effects of water relations and CO_2 assimilation on growth and yield of litchi trees.

Low yield of litchi is a major problem worldwide which is often attributed to improper monitoring of soil moisture supply and tree water relations. Poor fruit set, reduced fruit growth, sun-burning and fruit cracking are the major problems frequently faced by the growers. Formulation of a comprehensive guideline for the growers on this directions necessitated a better understanding of the dynamics of soil moisture content, tree-water status and their influence on growth, CO_2 assimilation, fruit retention, fruit development and productivity of litchi tree.

In this experiment, soil moisture was supplemented from anthesis through drip irrigation system at 80, 60, 40, 20 and 0% of pan coefficient (i.e., five treatments). The dynamics of soil moisture content and tree-water status were studied at weekly (in 2003) or fortnightly (in 2004) interval. The variations in shoot growth, net CO_2 assimilation rate, fruit set, retention and development of fruit and ultimate yield and quality of fruit at different irrigation levels were recorded.

4.3.1 Variation in soil water content at different irrigation levels

Variation in soil water content (SWC) in 2003 :The SWC was measured weekly at 50, 100 and 150 cm soil depths from anthesis (20.3.03)i.e., Week 1 to one week before harvest i.e., Week 9 (13.5.03). The initial SWC was estimated by gravimetric method for the calibration of neutron moisture probe. The variations in SWC (%v/v) due to treatments at different depths and weeks have been presented in Table 4.3.1.

Soil water content at 50 cm depth :The average SWC (average of five treatments) at 50 cm depth varied between 16.68% and 19.94% in different weeks (Table 4.3.1). In general, the average SWC declined gradually from 19.94% in week 1 to 16.68% in week 9, except in week 3 (19.46%) which showed slight increase over previous week (19.24%) as a result of light rainfall. However, the highest SWC (17.5% to 20.1%) was recorded by application of water at 80% pan coefficient (T_1) compared with the lowest (15.6% to 19.9%) in control (T_5: no-irrigation).

In week 1, there were slight variations (19.9% to 20.1%) in SWC due to treatments. However, from week 2 onwards marked variations in SWC were recorded due to different irrigation levels. The SWC at week 2 varied between 19.1% in control and 19.5% in T_1 (0.8 E_0). In week 3, the SWC was recorded between 19.4% and 19.9% under irrigations between 40% and 80% pan coefficient (T_3, T_2 and T_1) compared with 19.0% to 19.3% due to T_4 (0.2 E_0) and T_5 (control) treatments. The SWC showed sharp decline at weeks 4, 5, 6 and 9 under all treatments. The depletion in SWC, however, was much more under T_3, T_4 and T_5 treatments compared with T_1 and T_2 (0.6 E_0) treatments.

In week 4, the SWC was less than 18.5% under T_3, T_4 and T_5 treatments compared with 18.7% and 19.0% due to T_2 and T_1 treatments, respectively. Between weeks 5 and 6, T_1 treatment caused SWC between 18.0 % and 18.5% compared with 16.9% to 17.4% in control. When irrigation was applied between 40% and 80% pan coefficient (T_3, T_2 and T_1) resulted in higher SWC (17.1% to 18.0%) during weeks 7, 8 and 9, which was much lower (15.6% to 16.9%) due to T_4 and T_5 treatments.

Soil water content at 100 cm depth : The average SWC (average of five treatments) at 100 cm soil depth varied between 17.28% and 14.24% at different weeks which declined gradually from 17.28% in week 1 to 14.24% in week 9 (Table 4.3.1). The depletion in SWC at 100 cm soil depth showed almost similar trends compared with that in 50 cm depth at different weeks and irrigation levels.

The SWC at weeks 1 and 2 varied between 17.1% and 17.5% and 16.4% and 16.9%, respectively due to different irrigation levels. Irrigation at 80% pan coefficient (T_1) caused 17.5%, 16.9% and 16.5% SWC at weeks 1, 2 and 3, respectively compared with 17.1%, 16.4% and 15.7%, respectively in control (T_5). Between weeks 4 and 7, the depletion of SWC was found more in T_5 (control) and T_4 (0.2 E_0) treatments compared with T_1, T_2 and T_3 treatments.

The SWC declined from 15.3% at week 4 to 13.5% at week 7 due to dry treatment (T_5) compared with 16.1% to 14.8%, respectively in T_1 treatment. During weeks 8 and 9, the SWC varied between 14.5% and 15.0% by irrigating between 60% and 80% pan coefficient (T_2 and T_1), which showed variations between 13.6% and 14.3% due to dry treatment (T_5) and irrigation at 20% pan coefficient (T_4).

Soil water content at 150 cm depth : The average SWC (average of five treatments) at 150 cm soil depth showed slight decrease from 24.32% to 24.0% between Weeks 1 and 3, which then depleted sharply after week 3 and reached 19.9% in week 9 (Table 4.3.1). The different treatments showed very little variations in SWC during the first three weeks compared with the following weeks, indicating that extraction of soil moisture by tree had progressed from upper to lower soil layers gradually.

The highest SWC (21.7% to 24.4%) was recorded by irrigating at 80% pan coefficient compared with the lowest (17.4% to 24.3%) in control (T_5). Supplementing irrigation between 40% and 80% pan coefficient maintained SWC above 23.5% and 22.3% at weeks 4 and 5, respectively compared with below 23.5% at week 4 and below 22.3% at week 5 due to irrigation at 20% pan coefficient (T_4) and dry treatment (T_5). The depletion of SWC increased sharply between weeks 6 and 9 in T_4 and T_5 treatments. During weeks 6 to 9, the SWC varied between 21.1% and 22.7% in T_1 (0.8 E_0) compared with 17.4% and 21.1% in control (T_5).

Variation in soil water content in 2004 : The SWC was measured at 50, 100 and 150 cm soil depths at 15 days interval from anthesis (18.3.04)i.e., week 1 to one week before harvest i.e., week 9 (13.5.04). The variations in SWC due to irrigation levels at different weeks and soil depths have been presented in Table 4.3.2.

Soil water content at 50 cm depth : The average SWC (average of five treatments) varied between 17.4% and 20.7% in different weeks. Irrigation

104

at 80% pan coefficient (T_1) caused the maximum SWC of 18.8% to 23.4% compared with 15.1% to 19.0% in control (T_5). The SWC decreased gradually between weeks 1 and 3 under different treatments, which increased in weeks 5 and 7. Between weeks 5 and 7, the SWC varied between 21.8% and 23.4% in T_1 (0.8 E_0) compared with 17.4 and 19.0% in T_5 treatment. The SWC decreased from 19.0% in week 7 to 15.1% in week 9 under control (T_5) while it ranged between 20.2% and 23.4% by irrigating at 60% and 80% pan coefficient (T_2 and T_1).

Soil water content at 100 cm depth : The average SWC (average of five treatments) declined from 16.18% in week 1 to 15.7% in week 3, which however, increased between weeks 5 and 7 and subsequently declined to 17.3% in week 9 (Table 4.3.2). Different irrigation levels showed marked variations in SWC in different weeks of observation. Irrigation at 80% pan coefficient (T_1) caused the highest SWC (16.6% to 21.0%) compared with 14.6% to 16.0% in control (T_5). The different levels of irrigation caused significant variations in SWC at week 5 (15.2% to 18.8%), week 7 (16.0% to 21.0%) and week 9 (15.6% to 19.1%) compared with week 1 (15.8% to 16.7%) and week 3 (14.6% to 16.6%).

Soil water content at 150 cm depth :The average SWC at 150 cm soil depth (average of five treatments) varied between 23.96% and 24.64% in different weeks (Table 4.3.2). Different irrigation levels showed marked variations in SWC at week 5 (23.8% to 25.3%), week 7 (23.2% to 26.1%) and week 9 (17.8% to 25.4%) compared with week 1 (24.1% to 24.3%) and week 3 (23.4% to 24.2%). Maximum SWC of 24.2% to 26.1% was recorded by irrigating at 80% pan coefficient (T_1), while it was recorded lowest (17.8% to 24.1%) in control (T_5).

Variations in soil water content at 0-150 cm soil profile (average of two years) : The average SWC from 0 to 150 cm soil profile over two years period showed that the initial SWC at 0-150 cm soil profile was 19.87% which varied between 15.87% and 20.93% in the following weeks due to different levels of irrigation (Fig. 4.3.1). The levels of irrigation caused marked variations in SWC at 0-150 cm soil profile in all the weeks. However, the variations in SWC were more pronounced at week 9 (15.87% to 20.00%), week 7 (18.10% to 20.93%), week 5 (18.32% to 20.43%) and week 3 (18.77% to 20.03%) compared with week 1 (19.77% to 20.30%).

Application of irrigation at 80% pan coefficient (T_1) showed the highest SWC (20.00% to 20.93%) followed by 19.00% to 20.10% due to irrigation at 60% pan coefficient (T_2) in all weeks. The lowest SWC (15.87% to 19.77%) was however, recorded due to dry treatment (T_5). In other words, there was the minimum fluctuations in SWC which was almost static at 93.30% of field capacity (F.C.) in all the weeks due to irrigation at 80% pan coefficient (T_1) compared with the maximum fluctuations in SWC (90.7% of F.C. in week 1 to 72.8% of field capacity in week 9) due to dry treatment (T_5).

Irrigating between 20% and 40% pan coefficient (T_4 and T_3) however, showed 17.25% to 19.98% SWC in different weeks. At lower irrigation levels, the SWC fell below 90% of field capacity at week 3 (T_3, T_4 and T_5), week 5 (T_3, T_4 and T_5), week 7 (T_3, T_4 and T_5) and week 9 (T_2, T_3, T_4 and T_5).

4.3.2 Variation in leaf water potential due to different levels of irrigation levels

Variation in leaf water potential in 2003 : The morning leaf water potential (LWP) was measured from week 1 (20.3.03) to week 9 (13.5.03) and afternoon LWP was measured from week 5 (17.4.03) to week 9 (13.5.03) at weekly interval.

Morning LWP : The initial (12.3.03) LWP was -0.75 Mpa. The average of morning LWP (average of five treatments) showed gradual decline from -0.17 Mpa in week 1 to -1.69 Mpa in week 9, except in week 6 (-1.00 Mpa) and week 7 (-0.70 Mpa). Different irrigation levels showed significant variations in morning LWP in all the weeks, except in weeks 1 and 7 (Table 4.3.3). The morning LWP varied between -0.98 and -0.10 Mpa by irrigating at 80% pan coefficient (T_1) compared with -2.40 to -0.18 Mpa in dry treatment (T_5).

At lower irrigation levels, the morning LWP fell below -1.30 Mpa at week 3 (T_3, T_4 and T_5), week 4 (T_4 and T_5), week 5 (T_4 and T_5), week 6 (T_5), week 8 (T_3, T_4 and T_5) and week 9 (T_3, T_4 and T_5) compared with above -1.30 Mpa by irrigating at 60% and 80% pan coefficient (T_2 and T_1). The variations in LWP in the morning due to different irrigation levels was higher at week 9 (-2.40 to -0.98 Mpa), week 8 (-2.11 to -0.94 Mpa), week 6 (-1.71 to -0.60 Mpa), week 5 (-1.66 to -0.80 Mpa), week 4 (-1.92 to -0.96 Mpa), week 3 (-1.94 to -0.76 Mpa) and week 2 (-1.26 to -0.10 Mpa) compared with week 1 (-0.18 to -0.16 Mpa) and week 7 (-0.83 to -0.65 Mpa).

Afternoon LWP : The LWP in afternoon was always low (-2.30 to -1.30 Mpa) compared with that recorded in the morning LWP (-1.69 to -0.17 Mpa) at different weeks, irrespective of irrigation levels (Table 4.3.3). The average LWP in afternoon (average of five treatments) showed gradual decline from -2.02 Mpa in week 5 to -2.30 Mpa in week 9, except in week 7 (-1.30 Mpa) and week 8 (-2.03 Mpa). Different irrigation levels showed significant variations in afternoon LWP which varied between -2.76 and -1.16 Mpa due to different levels of irrigation. The T_1 treatment (irrigation at 80% pan coefficient) caused higher LWP in afternoon (-1.82 to -1.16 Mpa) compared with -2.76 Mpa to -1.64 Mpa in control (T_5:dry treatment). At lower irrigation levels, the afternoon LWP fell below -2.0 Mpa at week 5 (T_3, T_4 and T_5), week 6 (T_3, T_4 and T_5), week 8 (T_3, T_4 and T_5) and week 9 (T_3, T_4 and T_5).

Variation in leaf water potential in 2004 : The morning leaf water potential (LWP) was measured from week 1 (18.3.04) to week 9 (13.5.04) and afternoon LWP was measured from week 5 (15.4.04) to week 9 (13.5.04) at 15 days interval.

Morning LWP : The initial (5.3.03) LWP was -0.64 Mpa. The average morning LWP (average of five treatments) showed gradual decline from -0.86 Mpa in week 1 to -1.35 Mpa in week 9, except in week 3 (-1.34 Mpa). The morning LWP showed significant variations at different weeks due to different irrigation levels (treatments) which varied between -1.84 Mpa and -0.68 Mpa (Table 4.3.4).

Irrigation at 80% pan coefficient resulted the highest (-0.84 to -0.68 Mpa) LWP in the morning compared with the lowest (-1.84 to -1.10 Mpa) in dry treatment (T_5). At lower irrigation levels, the morning LWP fell below -1.30 Mpa at week 3 (T_3, T_4 and T_5), week 5 (T_4 and T_5), week 7 (T_3, T_4 and T_5) and week 9 (T_3, T_4 and T_5) compared with above -1.30 Mpa when irrigating at 60% and 80% pan coefficient (T_2 and T_1).

Afternoon LWP : The LWP in the afternoon was lower (-2.16 Mpa to -1.79 Mpa) compared with that recorded in the morning (-1.34 Mpa to -0.86 Mpa) at different weeks, irrespective of irrigation levels. The average LWP in the afternoon (average of five treatments) showed gradual decline from -1.79 Mpa in week 5 to -2.16 Mpa in week 9 (Table 4.3.4). Different irrigation levels showed significant variations in afternoon LWP which varied between -2.78 Mpa and -1.23 Mpa due to levels of irrigation. The T_1 treatment (irrigation at 80% pan coefficient) caused higher LWP in the afternoon (-1.44 Mpa to -1.23 Mpa) compared with afternoon LWP of -2.78 Mpa to -2.36 Mpa in control (T_5 : dry treatment). At lower irrigation levels, the afternoon LWP fell below -2.0 Mpa at week 5 (T_4 and T_5), week 7 (T_4 and T_5), week 8 (T_3, T_4 and T_5) and week 9 (T_3, T_4 and T_5).

Variation in morning leaf water potential (average of two years) : The average LWP in the morning in two years period showed variations between -2.02 Mpa and -0.72 Mpa due to different irrigation levels in different weeks (Fig. 4.3.2). In general, the initial leaf water status (LWP : -0.75 Mpa to -0.66 Mpa) in the morning gradually declined in the following weeks until it reached -2.02 Mpa to -0.88 Mpa in week 9, except in week 1 (-0.64 to 0.42 Mpa) and week 7 (-1.32 Mpa to -0.70 Mpa).

Irrigation at 80% pan coefficient (T_1) caused the highest (-0.88 Mpa to -0.42 Mpa) LWP in the morning compared with -2.02 Mpa to -0.64 Mpa in control (T_5). At lower irrigation levels, the morning LWP fell below -1.30 Mpa at week 3 (T_3, T_4 and T_5), week 5 (T_4 and T_5), week 7 (T_5)and week 9 (T_3, T_4 and T_5). But the morning LWP was above -1.30 Mpa in all the weeks due to irrigation at 60% and 80% pan coefficient (T_2 and T_1).

4.3.3 Variation in relative water content (RWC) of leaf due to irrigation levels

RWC of leaf in 2003 : The RWC of leaf was recorded at weekly interval from week 1 (20.3.03) to week 9 (13.5.03). The initial (12.3.03) RWC of leaf was 95.24%. The average RWC of leaf (average of five treatments) showed gradual decline from 94.38% in week 1 to 86.70% in week 6 which then slightly increased to 88.43% to 89.35% between weeks 7 and 9. Different irrigation levels showed significant variations in RWC of leaf in all the weeks under study, except in weeks 1 and 8 (Table 4.3.5).

Irrigation at 80% pan coefficient (T_1) caused the highest RWC of leaf (91.04% to 95.15%) compared with the lowest (82.11% to 93.78%) in dry treatment (T_5). At lower irrigation levels, the RWC of leaf fell below 90% at week 2 (T_5), week 3 (T_3 and T_5), week 4 (T_3, T_4 and T_5), week 5 (T_3, T_4 and T_5), week 6 (T_2, T_3, T_4 and T_5), week 7 (T_3, T_4 and T_5), week 8 (T_4 and T_5) and week 9 (T_3, T_4 and T_5).

RWC of leaf in 2004 : The RWC of leaf was recorded at 15 days interval from week 1 (18.3.04) to week 9 (13.5.04). The initial (5.3.04) RWC of leaf was 95.06%. The average RWC of leaf (average of five treatments) showed gradual decline from 90.38% in week 1 to 86.50% in week 9, however, it remained almost static (87.29% to 87.93%) between weeks 3 and 7.

Different irrigation levels also showed significant variations in RWC of leaf in all the weeks (Table 4.3.6).

Irrigation at 80% pan coefficient (T_1) resulted the highest RWC of leaf (91.20% to 93.55%) compared with the lowest (82.15% to 87.06%) due to dry treatment (T_5). At lower irrigation levels, the RWC of leaf fell below 90% at week 1 (T_4, T_5), week 3 (T_3, T_4 and T_5), week 5 (T_3, T_4 and T_5), week 7 (T_3, T_4 and T_5), week 9 (T_3, T_4 and T_5).

RWC of leaf at different irrigation levels (average of two years) :The average RWC of leaf (average of two years) showed gradual decline from week1 to week 9 (83.25% to 91.74% in week 9), except in week 7 (84.25% to 92.13%). Different irrigation levels showed marked variations in RWC of leaf in all the weeks under study which varied between 83.25% and 94.35% (Fig. 4.3.3).

The highest RWC of leaf (91.74 to 94.35%) was recorded due to irrigation at 80% pan coefficient (T_1) compared with the lowest (83.25% to 90.42%) in dry treatment (T_5). At lower irrigation levels, the RWC of leaf fell below 90% at week 3 (T_3, T_4 and T_5), week 5 (T_3, T_4 and T_5), week 7 (T_3, T_4 and T_5) and week 9 (T_3, T_4 and T_5) compared with above 90% in all weeks due to irrigation at 60 and 80% pan coefficient (T_2 and T_1).

4.3.4 Weekly variation in shoot growth due to irrigation levels

Weekly shoot growth in 2003 :The initial data on growth for all the trees under the experiment was recorded. Weekly increase in shoot length of the tagged shoots was recorded on a fixed date of every week and the percentage increase over previous week was calculated. The data recorded for nine weeks period revealed that the growth increase was higher in the first week (20.3.03) which gradually declined in the following weeks and reached minimum (1.82% over previous week) at 7th week and again increased at weeks 8 and 9 (Table 4.3.7). The different treatments showed significant variations in increase of shoot length in first, eighth and ninth weeks.

The increase in growth (average of five treatments) was recorded maximum in week 8 (5.68 % increase over previous week) followed by week 9 (5.10% over the previous week) while it was least (1.82%) in the week 7. Among the different treatments, treatment T_1 caused the maximum increase of shoot growth followed by T_2, T_3, T_4 and the least in T_5 (dry treatment as control).

The total increase in shoot growth varied between 17.89% in control (T_5) and 38.67% in T_1 (0.8 E_0). At lower irrigation levels (T_3, T_4 and T_5), the total shoot growth was less than 30% compared with 35.00% and 38.67% due to irrigation at 60% and 80% pan coefficient (T_2 and T_1), respectively.

Weekly shoot growth in 2004 :In 2004, the shoot length of tagged shoots were recorded at 15 days interval and the percentage increase over previous week was calculated. The initial (5.3.04) length of shoot was 11.73 cm which showed significant variations in growth due to irrigation levels (treatments) at weeks 1, 3, 5 and 7 (Table 4.3.8).

The average increase in growth (average of five treatments) showed 5.45% increase in week 1 and 12.32% in week 3 which then declined gradually in the following weeks and showed only 1.38% increase in week 9. The total increase in shoot growth between week 1 and week 9 was maximum (41.23%) by irrigating at 80% pan coefficient, while it was minimum (16.08%) in dry treatment (T_5). The weekly growth rate of shoot was less than 3.6% over previous week at lower irrigation levels (T_3, T_4 and T_5) compared with more than 4.2% at higher irrigation levels (T_2 and T_1).

Variation in shoot growth at different irrigation levels (average of two years) : The average increase in shoot growth over two years period showed that the shoot growth (average of five treatments) increased from 4.70% in week 1 to 7.36% in week 3 which gradually declined in the following weeks and showed 2.25% increase in week 7. The shoot growth then increased again to 3.24% in week 9 (Fig. 4.3.4).

Different irrigation levels, however, showed marked variations in shoot growth in all the weeks. Irrigation at 80% pan coefficient (T_1) caused the highest (39.95%) increase in total shoot growth compared with the lowest

(16.98%) in dry treatment (T_5). The increase in growth due to irrigation levels showed higher variations at week 9 (0.78 to 3.93%), week 5 (1.12 to 4.65%) and week 3 (2.49 to 5.93%) compared with week 1 (4.49 to 5.05%) and week 7 (1.40 to 1.81%).

4.3.5 Net CO_2 assimilation and Stomatal conductance at different irrigation levels

Net CO_2 assimilation and stomatal conductance in 2003: The observation on net CO_2 assimilation rate(A) and stomatal conductance(g_s) were recorded weekly of a sunlit leaf in each replication between 9 and 10 h.

The average (average of five treatments) net CO_2 assimilation rate showed a gradual decline from 10.28 $\mu molm^{-2}s^{-1}$ at week 1 to 5.46 $\mu molm^{-2}s^{-1}$ at week 8, except 7.08 $\mu molm^{-2}s^{-1}$ at week 7. The stomatal conductance also showed a similar trend which was recorded 26.20 $milimolm^{-2}s^{-1}$ in week 1 declined to 17.13 $milimolm^{-2}s^{-1}$ in week 4 and then varied between 20.15 and 23.52 $milimolm^{-2}s^{-1}$ during weeks 5 and 8. Different irrigation levels caused significant variations in net CO_2 assimilation rate and stomatal conductance in all the weeks (Table 4.3.9).

Irrigating at 80% pan coefficient (T_1) showed the highest net CO_2 assimilation rate (6.50 to 11.60 $\mu molm^{-2}s^{-1}$) and stomatal conductance (24.83 to 36.73 $milimolm^{-2}s^{-1}$) compared with the lowest net CO_2 assimilation rate (4.00 to 9.70 $\mu molm^{-2}s^{-1}$) and stomatal conductance (9.70 to 18.56 $milimolm^{-2}s^{-1}$) in dry treatment (T_5).

At lower irrigation levels, the net CO_2 assimilation rate showed marked reductions in all the weeks. The net CO_2 assimilation rate was less than 10.00 $\mu molm^{-2}s^{-1}$ in T_4 and T_5 treatments in week 1 and T_3, T_4 and T_5 treatments in week 2. In week 3, it was less than 8.00 $\mu molm^{-2}s^{-1}$ due to T_3, T_4 and T_5 treatments. Between weeks 4 and 8, the net CO_2 assimilation rate fell below 6.00 $\mu molm^{-2}s^{-1}$ due to lower supply of irrigation at week 4 (T_4 and T_5), week 5 (T_3, T_4 and T_5), week 6 (T_3, T_4 and T_5), week 7 (T_5) and week 8 (T_3, T_4 and T_5) compared with 6.10 $\mu molm^{-2}s^{-1}$ to 7.80 $\mu molm^{-2}s^{-1}$ by irrigating at 60% to 80% pan coefficient (T_2 and T_1).

The stomatal conductance also markedly reduced at lower levels of irrigation in all the weeks. The stomatal conductance fell below 20.00 $milimolm^{-2}s^{-1}$ in dry treatment, T_5 (in all weeks), irrigation at 20% pan coefficient, T_4 (in all weeks, except 1 and 2) and irrigation at 40% pan coefficient, T_3 (in weeks 3 and 4) compared with above 20.00 $milimolm^{-2}s^{-1}$ by irrigating at 60% to 80% pan coefficient (T_2 and T_1).

Net CO_2 assimilation and stomatal conductance in 2004 : The average (average of five treatments) net CO_2 assimilation rate and stomatal conductance showed gradual decline from 10.00 $\mu molm^{-2}s^{-1}$ and 29.38 $milimolm^{-2}s^{-1}$, respectively in week 1 to 5.48 $\mu molm^{-2}s^{-1}$ and 21.89 $milimolm^{-2}s^{-1}$, respectively in week 9. The net CO_2 assimilation rate and stomatal conductance, however, varied significantly at different irrigation levels in all the weeks (Table 4.3.10).

The highest net CO_2 assimilation rate (7.70 to 12.80 $\mu molm^{-2}s^{-1}$) and stomatal conductance (26.70 to 38.40 $milimolm^{-2}s^{-1}$) were recorded by irrigating at 80% pan coefficient (T_1) compared with the lowest net CO_2 assimilation rate (4.10 to 7.50 $\mu molm^{-2}s^{-1}$) and stomatal conductance (18.50 to 23.56 $milimolm^{-2}s^{-1}$) in control (T_5: dry treatment). Lower irrigation levels markedly reduced the net CO_2 assimilation rate and stomatal conductance in all the weeks. The net CO_2 assimilation was less than 10.00 $\mu molm^{-2}s^{-1}$ in week 1 when water was applied below 60% pan coefficient (T_3, T_4 and T_5) and in week 3 by irrigating below 80% pan coefficient (T_2, T_3, T_4 and T_5). Between weeks 5 and 9, the net CO_2 assimilation rate fell below 6.00 $\mu molm^{-2}s^{-1}$ at lower levels of irrigation at week 5 (T_4 and T_5), week 7 (T_4 and T_5) and week 9 (T_3, T_4 and T_5) compared with 6.30 $\mu molm^{-2}s^{-1}$ to 8.40 $\mu molm^{-2}s^{-1}$ by irrigating at 60% to 80% pan coefficient (T_2 and T_1). Similarly, the stomatal conductance fell below 25 $milimolm^{-2}s^{-1}$ due to dry treatment, T_5 (in weeks 1, 5, 7 and 9), irrigation at 20% pan coefficient, T_4 (in weeks 3, 5, 7 and 9) and irrigation at 40% pan coefficient, T_3 (in weeks 5, 7 and 9) compared with above 25 $milimolm^{-2}s^{-1}$ in all the weeks when irrigating at 60% to 80% pan coefficient (T_2 and T_1).

Net CO_2 assimilation and stomatal conductance at different irrigation levels (average of two years) : The average net CO_2 assimilation over two years (average of five treatments) showed that the net CO_2 assimilation rate varied between 10.13 $\mu molm^{-2}s^{-1}$ in week 1 and 5.48 $\mu molm^{-2}s^{-1}$ in week 9 (Fig. 4.3.5). The weekly average of net CO_2 assimilation rate varied between 9.09 $\mu molm^{-2}s^{-1}$ in T_1 (0.8 E_0) and 5.78 $\mu molm^{-2}s^{-1}$ in control (T_5: dry treatments).

Different irrigation levels (treatments) caused marked variations in net CO_2 assimilation rate in all the weeks which varied between 4.20 $\mu mol\ m^{-2}s^{-1}$ and 11.90 $\mu molm^{-2}s^{-1}$. Irrigation at 80% pan coefficient (T_1) caused the highest (7.10 to 11.90 $\mu molm^{-2}s^{-1}$) net CO_2 assimilation rate compared with the lowest (4.20 to 8.60 $\mu molm^{-2}s^{-1}$) in control. The net CO_2 assimilation rate was less than 10.00 $\mu molm^{-2}s^{-1}$ in week 1 by irrigating below 60% pan coefficient (T_3, T_4 and T_5). In week 3 (post fruit set stage), the net CO_2 assimilation rate fell below 9.00 $\mu molm^{-2}s^{-1}$ in T_3, T_4 and T_5 treatments that corresponded with the SWC (0-150 cm depth) below 90% of field capacity and morning LWP below -1.30 Mpa. In between weeks 5 and 7, the net CO_2 assimilation rate was less than 7.00 $\mu molm^{-2}s^{-1}$ at lower irrigation levels (T_3, T_4 and T_5) compared with 7.25 $\mu molm^{-2}s^{-1}$ to 8.45 $\mu molm^{-2}s^{-1}$ due to irrigation at 60% to 80% pan coefficient (T_2 and T_1). In week 9, irrigating at 80% pan coefficient (T_1) showed 7.10 $\mu molm^{-2}s^{-1}$ net CO_2 assimilation rate while it was 4.20 $\mu molm^{-2}s^{-1}$ in dry treatment (T_5).

The average stomatal conductance (average of five treatments) showed gradual decline from 27.79 $milimolm^{-2}s^{-1}$ in week 1 to 21.02 $milimolm^{-2}s^{-1}$ in week 9, except 23.08 $milimolm^{-2}s^{-1}$ in week 7 (Fig. 4.3.6). The weekly average of stomatal conductance varied between 30.19 $milimolm^{-2}s^{-1}$ in T_1 (0.8 E_0) and 17.40 $milimolm^{-2}s^{-1}$ in control (T_5: dry treatments). The highest stomatal conductance (26.45 to 36.84 $milimolm^{-2}s^{-1}$) was recorded by irrigating at 80% pan coefficient (T_1) compared with the lowest (15.65 to

114

19.53 milimolm^{-2}s^{-1}) in dry treatment (T$_5$). In week 3 (post fruit set stage), stomatal conductance fell below 25.00 milimolm^{-2}s^{-1} in T$_3$, T$_4$ and T$_5$ treatments which corresponds with the SWC (0-150 cm depth) below 90% of field capacity and morning LWP below -1.30 Mpa.

In the following weeks, the stomatal conductance fell below 25 milimol m^{-2}s^{-1} corresponding with the SWC (0-150 cm depth) below 90% of field capacity and morning LWP below -1.30 Mpa due to dry treatment, T$_5$ (in weeks 5, 7 and 9), irrigation at 20% pan coefficient, T$_4$ (in weeks 5, 7 and 9), irrigation at 40% pan coefficient, T$_3$ (in weeks 5, 7 and 9) and irrigation at 60% pan coefficient, T$_2$ (week 9).

4.3.6 Flowering and initial fruit set at different irrigation levels

Length of panicle : The length of panicle varied between 21.87 cm and 24.54 cm in 2003 compared with 26.90 cm and 29.46 cm in 2004 (Tables 4.3.11 and 4.3.12). Different irrigation levels showed significant effect on panicle length in 2004 and the length of panicle was more than 28 cm due to irrigation at or above 40% pan coefficient (T$_1$, T$_2$ and T$_3$). Irrigation at 80% pan coefficient caused maximum (29.46 cm) length of panicle in 2004, while it was minimum (26.90 cm) in dry treatment (T$_5$).

Breadth of panicle: The effect of different irrigation levels on breadth of panicle was found significant in 2004 but was not significant in 2003 (Tables 4.3.11 and 4.3.12). The average breadth of panicle (average of five treatments) was 13.39 cm in 2003 and 15.44 cm in 2004. Maximum (16.35 cm) breadth of panicle was recorded in 2004 when irrigation was applied at 80% pan coefficient (T$_1$) compared with 14.82 cm in control (T$_5$).

Total number of flowers per panicle : The average number of flowers per panicle was recorded higher (1566.57) in 2004 compared with 1382.38 per panicle in 2003. Total number of flowers per panicle showed significant variations in 2004 but was not significant in 2003 due to different treatments (Tables 4.3.11 and 4.3.12). Maximum (1806.71) number of flowers per panicle was recorded due to T$_1$ (0.8 E$_0$) treatment compared with 1281.36 per panicle in dry treatment (T$_5$).

115

Number of staminate and pistillate flowers produced per panicle : The average (average of five treatments) number of staminate and pistillate flowers per panicle were 994.07 and 338.31, respectively in 2003 and 1155.22 and 413.12, respectively in 2004 (Tables 4.3.11 and 4.3.12). The variations in staminate and pistillate flower production in both the years were however, not significant due to levels of irrigation. There were considerable variations in number of staminate and pistillate flowers per panicle at different irrigation levels. Those variations however, showed no specific trend with different irrigation levels in 2003, but the number of both types of flowers per panicle in 2004 showed gradual decline with the decrease in irrigation levels.

Sex ratio : The ratio between the number of staminate and pistillate flowers on a panicle varied between 2.53 : 1 and 2.74 : 1 in 2003 and between 2.59 : 1 and 3.01 : 1 in 2004, but showed no specific trends of variations with irrigation levels (Tables 4.3.11 and 4.3.12).

Initial Fruit set per panicle : The average number of fruitlets per panicle, recorded one week after anthesis, was much higher (17.78/panicle) in 2004 compared with 14.15 per panicle in 2003 (Tables 4.3.11 and 4.3.12). The number of fruits initially set per panicle showed gradual decline with the decrease in the levels of irrigation treatments in both the years (Fig. 4.3.7). In 2003, maximum (14.65) fruit set per panicle was recorded by irrigating at 80% pan coefficient (T_1) compared with 13.26 fruits set per panicle in dry treatment (T_5). The levels of irrigations showed significant effect on the fruit set per panicle in 2004. Irrigation at 80% pan coefficient caused maximum fruit set of 20.16 per panicle compared with 15.01 per panicle in dry treatment.

4.3.7 Weekly fruit drop at different irrigation levels

The average number of fruits retained per panicle (average of ten panicles per tree) was recorded weekly and weekly fruit drop (%) was calculated.

Weekly fruit drop in 2003 : The average (average of five treatments) fruit drop was 14.01% at one week after fruit set (WAFS) which reached maximum (21.35%) at 2 WAFS and then declined sharply in the following weeks (Table 4.3.13). Different irrigation levels showed significant effect on fruit drop at 2 WAFS and retention of fruit per panicle at harvest. At 2 WAFS, fruit drop reduced significantly due to higher levels of irrigation over dry treatment (33.64%). The number of fruits retained per panicle at harvest was found significantly higher (10.55 to 11.48/panicle) when irrigation was supplied at 60% or 80% pan coefficient (T_2 and T_1) compared with 6.07 to 9.24 per panicle with irrigation at or below 40% pan coefficient (T_3, T_4 and T_5). The total fruit drop was maximum (54.22%) due to dry treatment and minimum with irrigation at 80% pan coefficient.

Weekly fruit drop in 2004 : Different irrigation levels showed significant effect on weekly fruit drop in the first four weeks after fruit set and the number of fruits retained per panicle at harvest (Table 4.3.14). The fruit drop (average of five treatments) was recorded maximum (18.23%) at 2 WAFS compared with 11.50% at 1 WAFS and which then declined gradually from 3.03% at 3 WAFS to 0.21% at 7 WAFS. The number of fruits retained per panicle at harvest was significantly higher (16.03/panicle) by application of irrigation at 80% pan coefficient (T_5) compared with 7.35 per panicle in dry treatment (T_5). The total drop of fruit reduced gradually from 51.30% in dry treatment to 20.50% due to highest level of irrigation (T_1:0.8 E_0).

The average of weekly fruit drop due to different irrigation levels over two years showed that the period between 1st and 3rd WAFS was very much sensitive to moisture stress (Fig. 4.3.8). During this period, lower levels of irrigation (at or below 40% pan coefficient) as well as dry treatment caused more than 12.5% fruit drop at 1 WAFS and more than 18.0% fruit drop at 2 WAFS. A second peak of fruit drop was noticed at 4 WAFS in dry and semi-dry treatments (T_5, T_4 and T_3), which was however, absent in T_1 and T_2. The total drop of fruit was recorded lowest (21.07%) due to highest level of irrigation (T_1:0.8 E_0) treatment, while the dry treatment caused the maximum (52.76%) dropping of fruit (Fig. 4.3.7). There was a gradual decline in total drop of fruit with increase in levels of applied irrigation

117

treatments. Irrigation at or below 40% pan coefficient caused more than 35% fruit drop which reduced to less than 26% due to irrigation at or above 60% pan coefficient.

4.3.8 Weekly variation in peel, seed, aril and fresh weight of fruit

Fruit growth in 2003 :Fruit samples were collected weekly from 4th week after fruit set (22.4.03) for recording peel, seed, aril and fresh weight of fruit. At 4 WAFS, the average weight of fresh fruit and aril were 6.86 g and 1.11 g, respectively causing an aril content of 16.18% which increased to more than double (36.01%) at 5 WAFS and triple (53.68%) at 6 WAFS (Table 4.3.15). Different irrigation levels showed significant effect on fruit weight at 4, 5, 6 and 7 WAFS, while aril content of fruit varied significant due to irrigation levels at 5, 6 and 7 WAFS. The weight of peel and seed, however, did not show any significant variations due to levels of irrigation treatments. During the fruit development period (4, 5, 6 and 7 WAFS), irrigation applied at 80% pan coefficient caused maximum fruit weight of 7.27, 12.87, 17.88 and 24.95g, respectively and aril content of 18.16, 37.80, 55.54 and 64.37%, respectively compared with the minimum fruit weight of 6.35, 11.80, 16.58 and 21.27g, respectively and aril content of 14.02, 34.58, 51.98 and 57.41%, respectively in dry treatment.

Fruit growth in 2004 : The average weight of fruit and aril content at 4 WAFS were 6.75g and 1.05g, respectively causing an aril content of 15.50% which became more than double (33.06%)at 5 WAFS and triple (53.29%) at 6 WAFS (Table 4.3.16). The weight of fruit and aril content varied significantly due to different levels of irrigation during the fruit growth and development period. The aril weight at 4 WAFS and the weight of peel and seed at all the weeks, however, showed no significant variations due to different levels of irrigation. During fruit development period (4, 5, 6 and 7 WAFS), irrigation at 80% pan coefficient caused the maximum weight of fruit (7.02, 12.71, 18.31 and 24.71g, respectively) and aril content (17.52, 36.61, 55.62 and 65.21%, respectively) compared with the minimum fruit weight (6.25, 11.45, 16.10 and 22.23g, respectively) aril content (13.95, 30.75, 51.35 and 60.40%, respectively) in dry treatment.

4.3.9 Sun-burning and fruit-cracking at different irrigation levels

Moisture stress showed significant influence on the intensity of sun-burning and skin-cracking of fruit. The problem of sun-burning was noticed as early as 5 weeks after fruit set and the maximum number of sun-burnt fruits (12.95%) were recorded under no-irrigation (control) treatment at harvest (Table 4.3.17). The minimum intensity of sun-burning and skin-cracking were recorded by irrigating at 80% pan coefficient (3.64% and 2.20%, respectively), while it was as high as 12.95% and 9.67%, respectively in dry treatment (T_5) (Figs. 4.3.9 and 4.3.10).

The number of normal fruits at harvest significantly reduced due to moisture stress. Maximum (94.16%) number of normal fruits were harvested when trees were irrigated at 80% pan coefficient (T_1) compared with the minimum (77.38%) in dry treatment (T_5). Irrigation at or below 40% pan coefficient (T_3, T_4 and T_5) caused wastage of 11.52% to 22.62% of fruits at harvest due to sun-burning and skin-cracking. More than 93% of the harvested fruits were normal by application of irrigation during fruit growth and development period at 60% and 80% pan coefficient (T_2 and T_1).

4.3.10 Physical characters of fruit at different irrigation levels

Among the physical characters of fruit, the fruit weight and pulp weight showed significant variations in both the years of experimentation due to different levels of irrigation (Tables 4.3.18 and 4.3.19, Fig. 4.3.11).

Length and diameter of fruit : The length and diameter of fruit varied between 3.93 cm and 4.24 cm and 3.31cm and 3.47 cm, respectively in 2003 due to treatments. In 2004, it varied between 4.02 cm and 4.25 cm and 3.33 cm and 3.46 cm, respectively (Tables 4.3.18 and 4.3.19). The effect of different irrigation levels on length and diameter of fruit were not statistically significant. However, the length (4.18 to 4.25 cm) and diameter (3.44 to 3.47 cm) of fruit that were developed due to irrigation at 80% pan coefficient, were more than the average (average of five treatments) length (4.12 cm in 2003 and 4.13 cm in 2004) and diameter (3.14 cm in both years) of fruit in both the years.

*Fruit weight :*Irrigation at 80% and 60% pan coefficient (T_1 and T_2) significantly increased the weight of fruit over the dry treatment (T_5) in both the years (Tables 4.3.18 and 4.3.19). The maximum weight (24.95g) of fruit was recorded by irrigating at 80% pan coefficient (T_1) in 2003. The weight of fruit was minimum (21.27 to 22.23g) in both the years in dry treatment (T_5) compared with 24.71g to 24.95g in T_1 (0.8 E_0) treatment. A gradual decline in fruit weight was recorded due to increasing intensity of moisture stress. In limited supply of moisture due to irrigation at or below 40% pan coefficient, the fruit weight reduced below 23.70g in both the years (Fig. 4.3.11).

Aril content of fruit : Different irrigation levels showed significant effect on aril weight of fruit in both the years. The aril weight of fruit varied between 12.66g and 16.06g in 2003 and 13.43g and 16.01g in 2004, due to different levels of irrigation (Tables 4.3.18 and 4.3.19). Irrigating at 80% pan coefficient caused highest (64.37% in 2003 and 65.21% in 2004) aril content of fruit compared with the lowest (57.41% in 2003 and 60.86% in 2004) due to dry treatment. Moisture stress due to application of irrigation at or below 40% pan coefficient (T_3, T_4 and T_5) reduced the aril content (< 61.5%) of fruit that was significantly higher (> 63.5%) at irrigation levels of 60% and 80% pan coefficient (T_2 and T_1) (Fig. 4.3.11).

4.3.11 Influence of irrigation levels yield

Number of fruits per tree : Different irrigation levels showed significant variations in the number of fruits produced per tree in both the years (Tables 4.3.18 and 4.3.19). The number of fruits per tree varied between 1250.34 and 2711.48 in 2003 and between 1488.46 and 2981.60 in 2004. Higher levels of irrigation treatments (T_1 and T_2) significantly increased the number of fruits per tree in both the years. The highest (2981.60) number of fruits per tree was recorded in 2004 by irrigating at 80% pan coefficient (T_1) compared with 12.50.34 fruits per tree in 2003 in dry treatment (T_5). More than 2500 fruits per tree were produced on the trees that received irrigation at 60% and 80% pan coefficient but the yield markedly reduced to less than

2250 fruits per tree at lower levels (at or below 40%) of irrigation treatments.

Fruit yield per tree : The yield (kg/tree) varied significantly due to irrigation treatments in the years 2003 and 2004 (Tables 4.3.18 and 4.3.19). The average yield (average of five treatments) was 48.19 kg fruit per tree in 2003 and 51.78 kg fruit per tree in 2004. The highest yield of 67.65 kg fruit per tree in 2003 and 71.25 kg fruit per tree in 2004 were recorded when irrigation was applied at 80% pan coefficient (T_1) compared with 27.67 kg fruit per tree in 2003 and 32.60 kg fruit per tree in 2004 in dry treatment (T_5). The fruit yield increased gradually with increase in the levels of irrigation. Irrigation at 60% and 80% pan coefficient produced more than 65 kg fruit yield per tree which however, considerably reduced below 50 kg per tree due to irrigation at or below 40% pan coefficient (T_3, T_4 and T_5) (Fig. 4.3.12). The levels of irrigation between 20% and 80% of pan coefficient, increased fruit yield by 13.41% to 144.49% in 2003 and 14.11% to 118.56% in 2004 over the dry treatment.

4.3.12 Influence of irrigation levels on quality of fruit

Total Soluble Solids : Different levels of irrigation showed significant effect on Total Soluble Solids (TSS) content of fruit in both the years. The TSS content of fruit was recoded highest (18.87^0B in 2003 and 18.95^0B in 2004) at irrigation level of 80% pan coefficient (T_1) compared with the lowest (17.83^0B in 2003 and 17.95^0B in 2004) in dry treatment (T_5) (Tables 4.3.20 and 4.3.21). At lower irrigation levels (T_4 and T_5), considerable reduction was recorded in TSS content (17.83 to 18.30^0B) of fruit which were higher (18.60 to 18.95^0B) at higher levels of irrigation (T_2 and T_1) treatments.

Reducing sugars :The reducing sugar content of fruit varied between 14.15% and 15.12% in 2003 and 14.90% and 15.21% in 2004 due to different levels of irrigation (Tables 4.3.20 and 4.3.21). The maximum content of reducing sugar was recorded (15.12% in 2003 and 15.21% in 2004) due to irrigation at 60% pan coefficient (T_2) compared with 14.15%

in 2003 in T_4 (0.2 E_0) and 14.80% in 2004 in T_3 (0.4 E_0) treatments. The treatment effect on reducing sugar content of fruit was however, not significant.

Total sugars : The effect of different levels of irrigation treatments on total sugars content of fruit was found significant in 2003 and non-significant in 2004 (Tables 4.3.20 and 4.3.21). The total sugars content of fruit varied between 16.57% and 17.33% in 2003 and 16.70% and 17.10% in 2004 due to treatments. The highest amount of total sugars in fruit was recorded 17.33% in 2003 due to T_4 (0.2 E_0) and 17.10% in 2004 due to T_2 (0.6 E_0) treatments.

Tritratable acidity : The average (average of five treatments) fruit acidity content was recorded 0.54% in 2003 and 0.49% in 2004. Irrigating at 80% pan coefficient caused the minimum (0.42% in 2003 and 0.45% in 2004) acidity content of fruit which was noted maximum due to T_5 (0.63%) and T_3 (0.53%) treatments in 2003 and 2004, respectively. However, the variations in fruit acidity due to different levels of irrigation was not significant (Tables 4.3.20 and 4.3.21).

Sugar/acid ratio : Different levels of irrigation treatments showed considerable variations in sugar/acid ratio of fruit in both the years (2003 & 2004). Irrigating at 80% pan coefficient caused the maximum (39.69 and 37.33 in 2003 and 2004, respectively) sugar/acid ratio of fruit compared with 26.94 in 2003 due to T_5 and 31.51 in 2004 due to T_3 treatments (Tables 4.3.20 and 4.3.21). The sugar/acid ratio of fruit was more than 35 when irrigation was applied at or above 60% pan coefficient, which fell below 32 due to lower levels of irrigation at or below 40% pan coefficient (T_3, T_4 and T_5).

TSS/acid ratio : The TSS/acid ratio of fruit varied markedly at different irrigation levels in both the years (Tables 4.3.20 and 4.3.21). The average (average of five treatments) TSS/acid ratio of fruit was 35.09 in 2003 and 37.83 in 2004. The maximum TSS/acid ratio of 44.92 in 2003 and 42.11 in 2004 were recorded when irrigation was applied at 80% pan coefficient compared with the lowest (28.31 in 2003 and 34.52 in 2004) in dry treatment (T_5) in both the years. An overall decrease in the TSS/acid ratio of fruit was found with the increasing intensity of moisture stress. A partial to complete moisture stress (T_3, T_4 and T_5) reduced the TSS/acid below 35 in 2003 and below 37 in 2004 (Fig. 4.3.12).

4.3.13 Water Use (Et), Crop Factor (Kc) and Water Expense Efficiency (WEE) at different irrigation levels

The water use and crop factor were estimated at weekly interval in 2003 and at two week (about 15 days) interval in 2004.

Weekly water use, crop factor and water expense efficiency in 2003 : The weekly water use (average of five treatments) during the fruit growth and development period varied between 9.04 mm and 36.55 mm (Table 4.3.22). Different irrigation levels showed considerable variations in weekly Et in different weeks. Irrigation at 80% pan coefficient (T_1) caused marked increase in weekly Et (10.56 mm to 26.56 mm) at weeks 2, 4, 6, 7 and 9 compared with 7.50 mm to 24.00 mm in dry treatment. The weekly Et and crop factor (Kc), in general showed gradual decline with the decrease in the levels of irrigation, except in weeks 1, 3 and 8. The Kc at 80% pan coefficient varied between 0.479 and 0.746 in weeks 2, 4, 5, 6, 7 and 9 compared with 0.326 and 0.674 due to dry treatment (T_5).

The total Et of litchi cv. Bombai between anthesis and harvest was 17.93 cm (average of five treatments) which varied between 17.13 cm and 19.27 cm due to variations in irrigation levels. Highest level of irrigation treatment (T_1:0.8 E_0) caused the maximum (19.27 cm) of total Et which gradually declined at lower levels of irrigation treatments. The water expense efficiency (WEE) was estimated maximum (619.80 kg/ha/cm) due to irrigation at 60% pan coefficient (T_1). There was a marked reduction in WEE due to irrigation at or below 40% pan coefficient (T_3, T_4 and T_5). Dry treatment (T_5) caused the minimum (282.68 kg/ha/cm) water expense efficiency.

Water use, crop factor and water expense efficiency in 2004 : The average Et (average of five treatments) in about 15 days between anthesis and harvest showed gradual increase from 20.45 mm in week1 to 57.06 mm in week 9 (Table 4.3.23). The data on Et and crop factor (Kc) in 2004, showed much clear trends of variation due to different levels of irrigation compared with that recorded in 2003. The Et in about 15 days was much higher (25.05 mm to 64.80 mm) at highest level of irrigation (T_1:0.8 E_0) compared with 16.90 mm to 49.70 mm in dry treatment (T_5). Similarly, the Kc varied between 0.50 and 0.87 when irrigation was applied at 80% pan coefficient compared with 0.34 to 0.73 in dry treatment. Both Et and Kc showed gradual decline with lower levels of irrigation treatments.

The total Et between anthesis and harvest varied between 17.33 cm in dry treatment and 22.97 cm in T_1 (0.8 E_1) treatment. There was a marked reduction in total Et at lower levels of irrigation treatments. The average (average of five treatments) water expense efficiency (WEE) was 441.59 kg/ha/cm which varied between 329.20 kg/ha/cm in dry treatment and 542.33 kg/ha/cm in T_1 (0.8 E_0). The WEE was maximum (560.56 kg/ha/cm) due to irrigation at 60% pan coefficient (T_2). The lower levels of irrigation treatments (at or below 40% pan coefficient, i.e., T_3, T_4 and T_5) caused a marked reduction (< 435 kg/ha/cm) in WEE.

Weekly water use (average of two years) : In 2004, the water use was estimated at an interval of about 15 days (2 weeks) which represented the cumulative water use of two succeeding weeks (i.e., weeks 2&3, 4&5, 6&7, 8&9) and the average crop factor for respective periods was estimated. These data were used to calculate the weekly water use and crop factor in 2004 and the average of weekly water use and crop factor over 2003 and 2004 were then obtained.

The average of weekly water use (Et) over two years showed that the weekly Et between 16.56 mm and 29.48 mm at different weeks and irrigation levels (Fig. 4.3.13). At week1, the average (average of five treatments) Et was 23.39 mm which declined gradually in the following weeks until 18.37 mm in week 7 and then increased to 26.87 mm in week 9.

Irrespective of levels of irrigation, the Et was 21.99±4.31 mm per week. Different irrigation levels caused marked variations in weekly Et which varied between 24.14±3.95 mm at the highest level of irrigation (T_1:0.8 E_0) and 20.17±3.75 mm in dry treatment (T_5). The weekly Et showed gradual decline with the decrease in the levels of irrigation.

Crop factor (average of two years) : The crop factor (Kc) (average of two years) varied between 0.33 and 0.59 during the fruit growth and development period. It was 0.55 (average of five treatments) at week1 (during anthesis and fruit set), which gradually declined in the following weeks and reached 0.38 at weeks 5 and 7 and further increased to 0.54 at week 9 (Fig. 4.3.14). At higher irrigation level of 80% pan coefficient (T_1), the Kc was higher (0.43 to 0.59) compared with 0.33 to 0.52 in dry treatment (T_5). The overall value of Kc between anthesis and harvest was 0.52 due to irrigation at 80% pan coefficient, which declined at lower irrigation levels (T_2, T_3 and T_4) and was lowest (0.44)in dry treatment (T_5).

Total water use and water expense efficiency (average of two years) : Irrespective of levels of irrigation, the total water use (Et) of 22 years Bombai litchi trees between anthesis and harvest was 190.50 mm

with a corresponding water expense efficiency (WEE) of 452.50 kg/ha/cm. The total Et was recorded maximum (211.20 mm) by irrigating at 80% pan coefficient (T_1) and minimum (172.30 mm) in dry treatment (T_5) (Fig. 4.3.15). However, the WEE was maximum (590.18 kg/ha/cm) at 60% pan coefficient (T_2), closely followed by 578.18 kg/ha/cm in T_1 (0.8 E_0) treatment. There was a sharp decline in WEE at lower irrigation levels (at or below 40% pan coefficient, T_3 and T_4) which was recorded lowest (305.94 kg/ha/cm) in dry treatment (T_5).

4.3.14 Economics of drip irrigation in litchi

The cost of installation of drip irrigation system including pump house for one hectare of litchi orchard was Rs.1,20,000.00 with estimated life-span of 10 years. The cost of cultivation of litchi including 10% of installation cost of drip system was Rs. 47,728.00 per hectare. The gross return from sale of

9.78 tonnes of litchi fruits (average of four irrigation levels) at the whole sale price rate (@ Rs. 25.00 per kg) was Rs. 2,44,500.00. Thus, the net return was Rs. 1,96,772.00 per hectare with a benefit : cost ratio of 4.12 : 1.

Irrigating at 80% pan coefficient caused net return of Rs. 2,61,272.00 per hectare with a benefit : cost ratio of 5.47 :1, while the net return was very low (Rs. 1,34,101.00 / ha) in dry treatment (T_5) resulting in a poor benefit : cost ratio of 1.81 :1.

The economics of drip irrigation in litchi has been presented in Tables 4.3.24 and 4.3.25.

4.3.15 DISCUSSION

In this experiment, soil moisture was supplemented from anthesis through drip irrigation system at 80, 60, 40, 20 and 0% of pan coefficient (i.e., five treatments). The data on soil water content, shoot growth, CO_2 assimilation rate, yield and quality of fruit revealed marked variations due to different levels of irrigation.

Different levels of irrigation caused marked variations in soil water content (SWC) at 0-150 cm soil profile in all the weeks. However, the variations in SWC were more pronounced at week 9 (15.87% to 20.00%), week 7 (18.10% to 20.93%), week 5 (18.32% to 20.43%) and week 3 (18.77% to 20.03%) compared with week 1 (19.77% to 20.30%). Application of irrigation at 80% pan coefficient (T_1) showed the highest SWC (20.00% to 20.93%) followed by 19.00% to 20.10% due to irrigation at 60% pan coefficient (T_2) in all weeks. The lowest SWC (15.87% to 19.77%) was however, recorded due to dry treatment (T_5). In other words, there was the minimum fluctuations in SWC which was almost static at 93.30% of field capacity (F.C.) in all the weeks due to irrigation at 80% pan coefficient (T_1) compared with the maximum fluctuations in SWC (90.7% of F.C. in week 1 to 72.8% of field capacity in week 9) due to dry treatment (T_5). Irrigating between 20% and 40% pan coefficient (T_4 and T_3) however, showed 17.25% to 19.98% SWC in different weeks. At lower irrigation levels, the SWC fell below 90% of field capacity at week 3 (T_3, T_4 and T_5), week 5 (T_3,

T_4 and T_5), week 7 (T_3, T_4 and T_5) and week 9 (T_2, T_3, T_4 and T_5). It was possible to maintain the available soil moisture at near field capacity (T_1 :0.8 E_0) through drip irrigation when operated for frequent application of small quantity of water which is almost equal to the depletion in soil water content. The use of drip irrigation was reported advantageous over surface irrigation method because maintenance of uniform moisture almost at field capacity in root zone was possible (Patel *et al*, 1999). Gowda and Gowda (1990) reviewed the research works carried out in India on irrigation methods and found that drip irrigation caused higher yields of several crops than other irrigation methods (unspecified) because the soil moisture level in the root zone was near field capacity throughout the crop growth period with the optimum level of water application through drip system varying between 60% and 100% of loss due to evaporation.

The initial leaf water potential (LWP) (-0.75 to -0.66 Mpa) in the morning gradually declined in the following weeks until it reached -2.02 to -0.88 Mpa in week 9, except in week 1 (-0.64 to 0.42 Mpa) and week 7 (-1.32 Mpa to -0.70 Mpa). Irrigation at 80% pan coefficient (T_1) caused the highest (-0.88 Mpa to -0.42 Mpa) LWP in the morning compared with -2.02 Mpa to -0.64 Mpa in control (T_5). At lower irrigation levels, the morning LWP fell below -1.30 Mpa at week 3 (T_3, T_4 and T_5), week 5 (T_4 and T_5), week 7 (T_5) and week 9 (T_3, T_4 and T_5). But the morning LWP was above -1.30 Mpa in all the weeks due to irrigation at 60% and 80% pan coefficient (T_2 and T_1). The gradual decline in morning LWP from week1 to week 9 was due to the influence of weather (Menzel and Simpson, 1986c) which showed gradual increase in temperature and vapour pressure during the period (Table 9b). However, the variation in LWP at different irrigation levels could be attributed to the osmotic adjustment in response to varying water status in soil (Syvertsen, 1985, Batten *et al*, 1992). Roe *et al*. (1995) also studied the effect of irrigation and dry treatments on LWP and reported that the LWP of well-watered 'Tai So' litchi tree was -0.8 Mpa which fell below -3.5 Mpa due to severe dry treatment resulting in wilting.

The LWP in the afternoon was lower (-2.04 to -1.37 Mpa) compared with that recorded in morning (-1.34 to -0.86) and this diurnal variation in LWP was supposed to be due to the changes in leaf-air water conductance by high degree of exposure of leaves to direct radiation in the afternoon (Menzel and Simpson, 1986c). The average LWP in the afternoon (average of five treatments) also showed gradual decline from -1.91 Mpa in week 5 to -2.23 Mpa in week 9. Different irrigation levels showed significant variations in afternoon LWP (average of two years) which varied between -2.77 Mpa and -1.19 Mpa due to levels of irrigation. The T_1 treatment (irrigation at 80% pan coefficient) caused higher LWP in the afternoon (-1.44 Mpa to -1.23 Mpa) compared with afternoon LWP of -2.78 Mpa to -2.36 Mpa in dry treatment (T_5). At lower irrigation levels, the afternoon LWP fell below -2.0 Mpa at week 5 (T_4 and T_5), week 7 (T_4 and T_5), week 8 (T_3, T_4 and T_5) and week 9 (T_3, T_4 and T_5). In a green house experiment with potted litchi plants of cv. Wai Chee, Chaikiattiyos et al. (1994) recorded that the midday LWP fell to -4.0 Mpa when the plants experienced severe drought and even wilted. Roe et al. (1995) recorded the LWP value of -0.8 Mpa in well-watered plants which declined to -3.5 Mpa in sandy soil and -4.5 Mpa in clay soil due to dry treatments. Under moisture stress, the amino acids mainly proline and amides like betaine increased and accumulated in the leaves contributing to osmotic adjustment as LWP fell. Proline increased very sharply (10 to 100-fold) which arises do novo from glutamic acid and ultimately probably from carbohydrates (Morgan, 1984, Turner and Jones, 1980).

The highest relative water content (RWC) of leaf (91.74 to 94.35%) was recorded due to irrigation at 80% pan coefficient (T_1) compared with the lowest (83.25% to 90.42%) in dry treatment (T_5). At lower irrigation levels, the RWC of leaf fell below 90% at week 3 (T_3, T_4 and T_5), week 5 (T_3, T_4 and T_5), week 7 (T_3, T_4 and T_5) and week 9 (T_3, T_4 and T_5) compared with above 90% in all weeks due to irrigation at 60% and 80% pan coefficient (T_2 and T_1), which may be due to the higher profile moisture contents in the soil (Fig. 13). The decline in RWC of leaf below 90% corresponded to the LWP below -1.30 Mpa when the levels of applied irrigation decreased at

128

different weeks. Neumann *et al*. (1974) studied the RWC and LWP of corn, sorghum, soybean and sunflower and found a significant relationship of RWC with LWP for all the crops studied.

Different irrigation levels showed marked variations in shoot growth in all the weeks. Irrigation at 80% pan coefficient (T_1) caused the highest (39.95%) increase in total shoot growth compared with the lowest (16.98%) in dry treatment (T_5). The increase in growth due to irrigation levels, however, showed higher variations at week 9 (0.78 to 3.93%), week 5 (1.12 to 4.65%) and week 3 (2.49 to 5.93%) compared with week 1 (4.49 to 5.05%) and week 7 (1.40 to 1.81%). The effect of irrigation levels on growth of 'Mauritius' and 'Floridian' litchi was studied by Stern *et al*. (1998) in Israel. They observed that shoot growth was reduced significantly due to the lower levels (25% and 0% of class A pan evaporation coefficient) of irrigation. While better vegetative growth was obtained with highest level of irrigation (7.62 cm of water) compared to 2.54 cm of irrigation water (Singh and Pathak, 1983). Higher soil moisture regime due to irrigation at 80% pan coefficient (T_1) could have facilitated higher uptake of nutrients leading to increased shoot growth (Guimera *et al*., 1995).

The levels of irrigation (treatments) caused marked variations in stomatal conductance and net CO_2 assimilation rate in all the weeks which varied between 21.02 and 27.79 milimolm^{-2}s^{-1} and 4.20 and 11.90 µmolm^{-2}s^{-1}, respectively. Irrigation at 80% pan coefficient (T_1) caused the highest stomatal conductance (26.45 to 36.84 milimolm^{-2}s^{-1}) and net CO_2 assimilation rate (7.10 to 11.90 µmolm^{-2}s^{-1}) compared with the lowest stomatal conductance (15.65 to 19.53 milimolm^{-2}s^{-1}) and net CO_2 assimilation rate (4.20 to 8.60 µmolm^{-2}s^{-1}) in dry treatment (T_5). Similar observations were recorded by Menzel *et al*. (1995) in an experiment with 10 years old 'Tai So' litchi trees grown on a sandy loam soil. They recorded 70-300 milimolm^{-2}s^{-1} stomatal conductance and 3-13 µmolm^{-2}s^{-1} net CO_2 assimilation rate in well-watered treatment which reduced to 50-180 milimolm^{-2}s^{-1} stomatal conductance and 2-6 µmolm^{-2}s^{-1} net CO_2 assimilation rate due to dry treatment.

The net CO_2 assimilation rate was less than 10.00 $\mu molm^{-2}s^{-1}$ in week 1 by irrigating below 60% pan coefficient (T_3, T_4 and T_5). In week 3 (post fruit set stage), the stomatal conductance fell below 25.00 $milimolm^{-2}s^{-1}$ and the net CO_2 assimilation rate fell below 9.00 $\mu molm^{-2}s^{-1}$ due to low levels of irrigation (T_3, T_4 and T_5) treatments that corresponded with the SWC (0-150 cm depth) below 90% of field capacity and morning LWP below -1.30 Mpa. In the following weeks, the stomatal conductance fell below 25 $milimolm^{-2}s^{-1}$ corresponding with the SWC (0-150 cm depth) below 90% of field capacity and morning LWP below -1.30 Mpa due to dry treatment, T_5 (in weeks 5, 7 and 9), irrigation at 20% pan coefficient, T_4 (in weeks 5, 7 and 9), irrigation at 40% pan coefficient, T_3 (in weeks 5, 7 and 9) and irrigation at 60% pan coefficient, T_2 (week 9). In between weeks 5 and 7, the net CO_2 assimilation rate was less than 7.00 $\mu molm^{-2}s^{-1}$ at lower irrigation levels (T_3, T_4 and T_5) compared with 7.25 to 8.45 $\mu molm^{-2}s^{-1}$ due to irrigation at 60% to 80% pan coefficient (T_2 and T_1). In week 9, irrigating at 80% pan coefficient (T_1) showed 7.10 $\mu molm^{-2}s^{-1}$ net CO_2 assimilation rate while it was 4.20 $\mu molm^{-2}s^{-1}$ in dry treatment (T_5). The decline in stomatal conductance and net CO_2 assimilation rate as the leaf water potential fell (below -1.30 Mpa) under dry (T_5) and partial dry (T_4, T_3) treatments was also reported by Chaikiattiyos et al. (1994), Roe et al. (1995) and Menzel et al.(1995) with different experiment sites and varieties.

In subtropical Australia, Menzel et al. (1995) recorded 70-300 $milimolm^{-2}s^{-1}$ stomatal conductance and 3-13 $\mu molm^{-2}s^{-1}$ net CO_2 assimilation rate in well-watered 'Tai So' litchi tree which declined to 50-180 $milimolm^{-2}s^{-1}$ and 2-6 $\mu molm^{-2}s^{-1}$, respectively by withholding irrigation between late July and January. The net CO_2 assimilation rate of potted plants of cv. 'Wai Chee' declined as midday LWP fell below -1.0 Mpa and approached zero at a LWP of about -4.0 Mpa when plants wilted (Chaikiattiyos et al., 1994). Salisbary and Ross (2003a) proposed five possibilities to explain the responses of mesophytes (plants with characteristics intermediate between xerophytes and hydrophytes) to water stress : i) water activity (indicating its ability to enter into chemical reactions) is a function of water potential and is lowered by water stress, ii) solutes increase in concentration as water is lost, iii) water stress might result in special changes in membranes, iv) water stress might upset the hydration of macromolecules so that dehydration of key enzymes would cause disulfide bonds within proteins to break and reform, sometimes reforming between adjacent molecules, leading to enzyme denaturation when molecules are rehydrated and v) water stress may profoundly change the turgor pressure which could be the stimulus to which some special response mechanism in the cell reacts in transducing water stress to the observed cellular responses. Among the observed responses, decreased leaf area together with smaller stomatal conductance (g_s) are the major ways of slowing loss of water but, as a consequence, net CO_2 assimilation rate falls due to smaller leaf area per plant and decreased g_s and CO_2 entry. Photosynthesis is slowed by tissue dehydration initially by stomatal conductance. With longer exposure to, and greater degree of dehydration, particularly with decreased relative water content, low turgor and decreased osmotic potential which are internal to plasma membrane and therefore, likely to be the components of changed water balance perceived by cellular contents, mesophyll cells lose the ability to photosynthesize (Lawlor, 2001). The decline in net CO_2 assimilation rate due to water stress has been ascribed to several photosynthetic processes such as : i) water stress may decrease Rubisco activity (the enzyme that catalyzes the reaction of CO_2 fixation), ii) water stress may increase photorespiration resulting in lower net CO_2 assimilation rate, iii) water stress may decrease the ATP content of leaves and this is probably the cause of decreased content of Rubisco (RuBP) in chloroplast correlating with decreased photosynthesis, iv) the photosynthetic system may not tolerate the metabolic imbalances in leaf caused under stress condition due to a large ratio of NADPH to ATP resulting from the continued reduction of $NADP^+$ at a substantial rate even when CO_2 assimilation is impaired and v) in response to water stress, the accumulation of neutral osmotica such as proline may minimize exposure of the photosynthetic system to decreased osmotic potential and protect cells and membranes against increased concentration of ions.

In 2003, the treatments (levels of irrigation) were imposed in March when the growth and development of panicle was almost completed and naturally there was no significant effect of irrigation levels in 2003 on different flowering characters (length and breadth of panicle, total number of flowers per panicle and number of staminate and pistillate flowers per panicle). The flowering characters, however, showed significant variations in the following year (2004) due to different irrigation levels which indicate the influence of irrigation levels on the reproductive growth of litchi. Application of irrigation at 60% and 80% pan coefficient during fruit growth and development period in 2003 influenced the trees for uptake of more nutrients which increased flowering in 2004. Maximum length (29.46 cm) and breadth (16.35 cm) of panicle in 2004 were recorded by irrigating at 80% pan coefficient, while it was minimum (26.90 cm and 14.82 cm, respectively) in dry treatment. The number of staminate and pistillate flowers per panicle showed no specific trend with different irrigation levels in 2003, but the number of both types of flowers per panicle in 2004 showed gradual decline with decrease in the levels of irrigation. Hasan and Chattopadhyay (1991) recorded less number of total flowers per panicle, staminate and pistillate flowers per panicle by irrigating at 60% available soil moisture (ASM) depletion compared with irrigating at 30% and 45% ASM depletion.

The ratio of the number of staminate and pistillate flowers on a panicle varied between 2.53 : 1 and 2.74 : 1 in 2003 and between 2.59 : 1 and 3.01 : 1 in 2004, but showed no specific trend with irrigation levels. The number of fruits set per panicle showed gradual decrease at the lower levels of irrigation treatments in both the years. In 2003, maximum (14.65) fruits set per panicle was recorded by irrigating at 80% pan coefficient (T_1) compared with 13.26 fruits set per panicle in dry treatment (T_5). The different levels of irrigations showed significant effect on fruit set per panicle in 2004. Irrigation at 80% pan coefficient caused maximum fruit set (20.16/panicle) compared with 15.01 per panicle in dry treatment. Hasan and Chattopadhyay (1991) studied the effect of soil moisture regime on fruit set and observed minimum (18.24) fruit set per panicle in control (unirrigated) trees compared with 26.69 and 26.13 fruits per panicle by irrigating at 30% and 45% ASM depletion, respectively. The final set of fruit per panicle at harvest was recorded 2.0 ± 0.5 in dry treatment versus 7.0 ± 0.3 in well-watered treatment by Menzel et al. (1995a).

Data on weekly fruit drop reveals that the period between 1st and 3rd week after fruit set (WAFS) was very sensitive to moisture stress (Fig. 20). During this period, lower levels of irrigation (at or below 40% pan coefficient) as well as dry treatment caused more than 12.5% fruit drop at 1 WAFS and more than 18.0% fruit drop at 2 WAFS. Another increase in fruit drop was noticed at 4 WAFS in dry and at lower levels of irrigation treatments (T_5, T_4 and T_3), which was however, absent in T_1 and T_2 treatments. The total drop of fruit (until maturity) was lowest (21.07%) at the highest level of irrigation ($T_1:0.8 E_0$) treatment, while the dry treatment caused the maximum fruit drop of 52.76 percent. There was a gradual decline in total drop of fruit with increase in the levels of irrigation treatments. Irrigation at or below 40% pan coefficient caused more than 35% fruit drop which reduced to less than 26% due to irrigation at or above 60% pan coefficient. Menzel et al. (1995) reported that in 10 years old 'Tai So' litchi water deficits reduced initial fruit set by 30% and final fruit set by 70%. The decline in net CO_2 assimilation rate as LWP fell below -1.30 Mpa under dry (T_5) and partial dry (T_4, T_3) treatments appeared to have adverse effect on retention and growth and development of fruit, which is supposed to depends on current CO_2 assimilation (Roe et al., 1997).

The fruit and aril weight varied significantly due to different levels of irrigation. The weight of peel and seed at all the weeks, however, did not show any significant variations due to levels of irrigation. During the fruit development period (4, 5, 6 and 7 WAFS), irrigation at 80% pan coefficient caused maximum weight of fruit (7.14, 12.79, 18.09 and 24.83g, respectively) and aril content (17.84, 37.21, 55.58 and 64.79%, respectively) compared with minimum fruit weight (6.30, 11.63, 16.34 and 21.75g, respectively) and aril content (13.99, 32.67, 51.66 and 58.91%, respectively) in dry treatment. Dry treatment (T_5) thus reduced weight of fruit by 12.40%

and aril content of fruit by 9.07% compared with irrigating at 80% pan coefficient (T_1). Menzel *et al.* (1995) reported that water deficits reduced fruit growth (16.2±1.3g versus 23.0±0.3g in well-watered treatment) and proportion of aril in fruit (66.9±0.9% versus 72.0±0.6% in well-watered treatment) mainly due to a reduction in skin and aril weights.

Moisture stress showed significant influence on the intensity of sun-burning and skin-cracking of fruit. The minimum intensity of sun-burning and skin-cracking were recorded by irrigating at 80% pan coefficient (3.64% and 2.20%, respectively), while it was as high as 12.95% and 9.67%, respectively in dry treatment (T_5). Menzel *et al.* (1995) observed 30-40% more splitting of fruits due to dry treatment over well-watered treatment. Maximum number of normal fruits (94.16%) were harvested when trees were irrigated at 80% pan coefficient (T_1) compared with the minimum (77.38%) in dry treatment (T_5). Irrigation at or below 40% pan coefficient (T_3, T_4 and T_5) caused more loss of fruits at harvest due to sun-burning and skin-cracking incidence. More than 93% of the harvested fruits were normal by application of irrigation during fruit growth and development period at 60% and 80% pan coefficient (T_2 and T_1). Several factors have been reported to be associated with the phenomenon of fruit cracking. The cultivars with relatively thin shin, few tubercles per unit area, lower content of calcium and boron in pericarp were related to fruit cracking (Sanyal *et al.*, 1990, Kanwar *et al.*, 1972, Li *et al.*, 1995). Fruits on south-west side of the tree suffered more than north-east side due to shading. High temperature (>38°C) in combination with low humidity (RH<60%) favoured cracking (Kanwar and Nijjar, 1975). Inadequate moisture during fruit growth and development resulted skin becoming hard and inelastic (sunburnt), which is likely to crack when subjected to increased internal pressure as a result of rapid aril growth (Maiti and Mitra, 2001). Water stress, especially during rapid aril growth, reduced the calcium accumulation in fruit peel and thus increased the fruit cracking rate (Li *et al.*, 1999). In the present study, irrigation at 80% pan coefficient (T_1) maintained SWC at near field capacity and supposed to influence the associated factors of cracking. Adequate SWC is believed to have increased the uptake and accumulation of nutrients, especially calcium in the peel and thus reduced cracking (Li *et al.*, 1995, 1999). The plants avoids potential lethal dehydration by keeping its stomata less open or even closed in dry conditions (Daniel, 2002) but when the soil moisture is available the plant could continue transpiration which often plays an extremely important role in leaf cooling (Salisbury and Ross, 2003). Plants evaporate tremendous amounts of water into their environments, and each gram of water transpired absorbs 2.45 kJ (586 Cal) from the leaf and its environment. Such canopy (leaves plus fruit bunch) is often cooler than the surrounding air, even though it is in full sunlight (Salisbury and Ross, 2003b). Supplementing irrigation at 60% and 80% pan coefficient during fruit growth and development period, therefore, might have reduced the incidence of sun-burning and cracking of fruit over dry (T_5) and partial dry (T_4, T_3) treatments by increasing accumulation of calcium in peel and minimizing the detrimental effect of high temperature on the developing fruits.

The length (4.18 cm to 4.25 cm) and diameter (3.44 cm to 3.47 cm) of fruits that were developed by irrigating at 80% pan coefficient were more than the average (average of five treatments) length (4.12 cm in 2003 and 4.13 cm in 2004) and diameter (3.14 cm in both years) of fruits in both the years. Application of irrigation at 80% and 60% pan coefficient (T_1 and T_2) significantly increased the weight of fruit over the dry treatment (T_5) in both the years. The weight of fruit was minimum (21.27g to 22.23g) in both the years due to dry treatment (T_5) compared with 24.71 to 24.95g in T_1 (0.8 E_0) treatment. Different irrigation levels showed significant effect on pulp weight of fruit in both the years. Application of irrigation at 80% pan coefficient caused highest (64.37% in 2003 and 65.21% in 2004) aril content of fruit compared with the lowest (57.41% in 2003 and 60.86% in 2004) due to dry treatment. Moisture stress due to application of irrigation at or below 40% pan coefficient (T_3, T_4 and T_5) reduced the aril content (< 61.5%) of fruit which was significantly higher (> 63.5%) at irrigation levels of 60% and 80% pan coefficient (T_2 and T_1). Replenishment of potential water loss through drip irrigation resulted in increased size and weight of fruit in several fruit crops including litchi (Menzel *et al.*, 1995), sweet oranges (Kumar and Bojappa, 1994) and olive (Chartzoulakis *et al.*, 1992).

Different levels of irrigation significantly influenced the number of fruits per tree and fruit yield per tree in both years of experimentation. More than 2500 fruits per tree were produced on the trees that received irrigation at 60% and 80% pan coefficient however, the yield markedly reduced to less than 2250 fruits per tree at lower levels (at or below 40%) of irrigation. The highest yield of 67.65 kg fruit per tree in 2003 and 71.25 kg fruit per tree in 2004 were recorded by irrigating at 80% pan coefficient (T_1) compared with 27.67 kg fruit per tree in 2003 and 32.60 kg fruit per tree in 2004 in dry treatment (T_5). The levels of irrigation between 20% and 80% of pan coefficient, increased fruit yield by 13.76% to 131.52% over the dry treatment. Reduction in yield of litchi from 51.4±5.5 kg tree^{-1} in well-watered trees to 7.4±3.3 kg tree^{-1} in dry treatment was recorded by Menzel *et al.* (1995). The yield of several fruit crops under drip irrigation has been reportedly increased by 15.80% to 34.30% in banana, 30.60% in kinnow, 44.40% in lemon, 43.50% in papaya and 48.95% in pomegranate (Singh *et al.*, 2000).

Different levels of irrigation showed significant effect on total soluble solids (TSS) content of fruit in both the years. The TSS content of fruit was recorded highest (18.87^{0}B in 2003 and 18.95^{0}B in 2004) by irrigating at 80% pan coefficient (T_1) compared with the lowest (17.83^{0}B in 2003 and 17.95^{0}B in 2004) in dry treatment (T_5). The reducing sugar and tritratable acidity content of fruit were however, showed no significant effect in both years at different irrigation levels. The total sugar content of fruit showed significant variations in 2003 and non-significant in 2004 at different irrigation levels. Maximum content of reducing sugar was recorded (15.12% in 2003 and 15.21% in 2004) by irrigating at 60% pan coefficient (T_2) compared with 14.15% in 2003 in T_4 (0.2 E_0) and 14.80% in 2004 in T_3 (0.4 E_0) treatments. The highest amount total sugar content of fruit was recorded (17.33%) in 2003 due to T_4 (0.2 E_0) and 17.10% in 2004 due to T_2 (0.6 E_0) treatments. Irrigating at 80% pan coefficient caused the minimum (0.42% in

2003 and 0.45% in 2004) fruit acidity which was maximum in T_5 (0.63%) in 2003 and T_3 (0.53%) treatments in 2004. The sugar/acid ratio of fruit was more than 35 when irrigation was applied at or above 60% pan coefficient, which fell below 32 at lower levels of irrigation (T_3, T_4 and T_5). An overall decrease in the TSS/acid ratio of fruit was found with the increasing intensity of moisture stress. A partial to complete moisture stress (T_3, T_4 and T_5) reduced the TSS/acid ratio below 35 in 2003 and below 37 in 2004. The maximum TSS/acid ratio of 44.92 in 2003 and 42.11 in 2004 were recorded by irrigating at 80% pan coefficient. Kumar and Bojappa (1994) noted improvement in fruit quality of sweet orange using drip irrigation at 12 litre tree[-1] daily with 1 or 2 emitters. Higher fruit mineral content was related to irrigation of mango at shorter intervals (Wagner et al., 1984).

The weekly water use (Et) of litchi cv. Bombai varied between 16.56 mm and 29.48 mm at different weeks and levels of irrigation. At week1, the average Et was 23.39 mm which gradually declined in the following weeks until 18.37 mm in week 7 and it then increased to 26.87 mm in week 9. Irrespective of the levels of irrigation, the Et was 21.99±4.31 mm per week. In well-watered trees of 'Tai So' litchi grown on sandy loam soil in subtropical Australia, water use was recorded as 26±1 mm week[-1] (Menzel et al., 1995a). Different irrigation levels caused marked variations in weekly Et which varied between 24.14±3.95 mm in highest level of irrigation (T_1:0.8 E_0) and 20.17±3.75 mm in dry treatment (T_5). The weekly Et showed gradual decline at lower levels of irrigation treatments. Water use of papaya was found to decline with the decrease in evaporation-replenishment rates from 120% to 20% (Srinivas, 1996).

The crop factor (Kc) of litchi cv. Bombai (average of two years) varied between 0.33 and 0.59 during the fruit growth and development period. Menzel et al. (1995a) reported a crop factor of 0.4 to 1.2 between panicle emergence and four weeks after fruit harvest of 'Tai So' litchi in Australia. At higher irrigation level of 80% pan coefficient (T_1), the Kc was higher (0.43 to 0.59) compared with 0.33 to 0.52 in dry treatment (T_5). The overall value of Kc between anthesis and harvest was 0.52 by irrigating at 80% pan coefficient, which declined at lower levels (T_2, T_3 and T_4) of irrigation and was found lowest (0.44) in dry treatment (T_5).

Irrespective of the levels of irrigation, the total water use (Et) of 22 years Bombai litchi trees between anthesis and harvest was 190.50 mm with a corresponding water expense efficiency (WEE) of 452.50 kg/ha/cm. The total Et was recorded maximum (211.20 mm) by irrigating at 80% pan coefficient (T_1) and minimum (172.30 mm) in dry treatment (T_5). However, the WEE was maximum (590.18 kg/ha/cm) when irrigated at 60% pan coefficient (T_2), closely followed by 578.18 kg/ha/cm in T_1 (0.8 E_0) treatment. There was a sharp decline in WEE at lower irrigation levels (at or below 40% pan coefficient, T_3 and T_4) and it was lowest (305.94 kg/ha/cm) in dry treatment (T_5). Drip irrigation reportedly increased the water use efficiency of many fruit crops viz., banana, guava, grapes, kinnow, papaya, pomegranate, lemon and ber over surface method of irrigation (Singh *et al.*, 2000). The total water use of banana under basin irrigation method showed gradual increase from 924 mm at 20% evaporation-replenishment to 2372 mm at 120% evaporation-replenishment, while the water use efficiency was maximum (490 kg/ha-mm) at 60% evaporation-replenishment treatment (Hegde and Srinivas, 1990).

The benefit : cost ratio of litchi cultivation under drip irrigation (average of four irrigation levels) was 4.12 : 1. The economics of banana cultivation under drip irrigation was estimated by Hegde and Srinivas (1990) and the benefit : cost ratio was 3.73 : 1. In present experiment, irrigation at 80% pan coefficient (T_1) caused net return of Rs. 2,61,272.00 per hectare with a benefit : cost ratio of 5.47 : 1, while the net return was very low (Rs. 1,34,101.00) in dry treatment (T_5) resulting in a very poor benefit : cost ratio of 1.81 : 1.

Table 4.3.1(a) : Weekly variations in soil water content (%V) due to different levels of irrigation (2003)

Treatment	Initial (6.3)			Week 1 (20.3)			Week 2 (26.3)		
	50 cm	100 cm	150 cm	50 cm	100 cm	150 cm	50 cm	100 cm	150 cm
$T_1(0.8E_0)$				20.1	17.5	24.4	19.5	16.9	24.3
$T_2(0.6E_0)$				20.0	17.4	24.3	19.3	16.7	24.3
$T_3(0.4E_0)$	19.10	17.20	24.30	19.8	17.3	24.3	19.2	16.5	24.3
$T_4(0.2E_0)$				19.9	17.1	24.3	19.1	16.4	24.3
T_5(Control)				19.9	17.1	24.3	19.1	16.4	24.3
Average	-	-	-	19.9	17.3	24.3	19.2	16.6	24.3

Treatment	Week 3 (1.4)			Week 4 (7.4)			Week 5 (16.4)		
	50 cm	100 cm	150 cm	50 cm	100 cm	150 cm	50 cm	100 cm	150 cm
$T_1(0.8E_0)$	19.9	16.5	24.2	19.0	16.1	23.8	18.5	15.3	22.9
$T_2(0.6E_0)$	19.7	16.1	24.1	18.7	15.8	23.7	18.1	15.1	22.8
$T_3(0.4E_0)$	19.4	15.8	24.0	18.3	15.6	23.5	17.9	14.7	22.3
$T_4(0.2E_0)$	19.3	15.9	23.8	18.3	15.5	23.4	17.8	14.5	22.2
T_5(Control)	19.0	15.7	23.9	18.0	15.3	23.2	17.4	14.2	21.9
Average	19.46	16.0	24.0	18.46	15.66	23.52	17.94	14.76	22.42

Table 4.3.1 (b) : Weekly variations in soil water content (%V) due to different levels of irrigation (2003).

Treatment	Week 6 (22.4)			Week 7 (29.4)		
	50 cm	100 cm	150 cm	50 cm	100 cm	150 cm
$T_1(0.8E_0)$	18.0	15.0	22.7	17.9	14.8	22.4
$T_2(0.6E_0)$	17.5	14.5	22.7	17.2	14.5	22.5
$T_3(0.4E_0)$	17.3	14.1	22.1	17.1	14.4	21.8
$T_4(0.2E_0)$	17.2	14.0	21.7	16.9	14.0	21.1
$T_5(Control)$	16.9	13.8	21.1	16.3	13.5	20.6
Average	17.38	14.28	22.06	17.08	14.24	21.68

Treatment	Week 8 (6.5)			Week 9 (13.5)		
	50 cm	100 cm	150 cm	50 cm	100 cm	150 cm
$T_1(0.8E_0)$	18.0	15.0	22.1	17.5	14.9	21.7
$T_2(0.6E_0)$	17.5	14.6	21.8	17.3	14.5	21.0
$T_3(0.4E_0)$	17.2	14.4	21.1	17.1	14.3	19.8
$T_4(0.2E_0)$	16.1	14.3	20.8	15.9	14.0	19.6
$T_5(Control)$	15.9	13.6	19.8	15.6	13.7	17.4
Average	16.94	14.38	21.12	16.68	14.28	19.9

Table 4.3.2 : Weekly variations in soil water content (%V) due to different irrigation levels (2004)

Treatment	Initial (5.3)			Week 1 (18.3)		
	50 cm	100 cm	150 cm	50 cm	100 cm	150 cm
T_1 (0.8E_0)	18.30	16.10	24.20	18.8	16.7	24.3
T_2 (0.6E_0)				18.4	16.3	24.2
T_3 (0.4E_0)				18.2	16.1	24.2
T_4 (0.2E_0)				17.8	16.0	24.1
T_5 (Control)				17.4	15.8	24.1
Average				18.12	16.18	24.18

Treatment	Week 3 (1.4)			Week 5 (15.4)		
	50 cm	100 cm	150 cm	50 cm	100 cm	150 cm
T_1 (0.8E_0)	18.8	16.6	24.2	21.8	18.8	25.3
T_2 (0.6E_0)	18.2	16.0	24.2	20.4	17.4	24.2
T_3 (0.4E_0)	17.4	15.9	24.0	19.8	16.2	24.0
T_4 (0.2E_0)	16.6	15.4	24.0	18.1	16.0	24.0
T_5 (Control)	16.0	14.6	23.4	17.4	15.2	23.8
Average	17.4	15.7	23.96	19.5	16.72	24.26

Treatment	Week 7 (29.4)			Week 9 (13.5)		
	50 cm	100 cm	150 cm	50 cm	100 cm	150 cm
T_1 (0.8E_0)	23.4	21.0	26.1	21.4	19.1	25.4
T_2 (0.6E_0)	21.2	19.3	25.6	20.2	18.4	22.6
T_3 (0.4E_0)	20.8	18.4	24.2	18.5	16.8	22.4
T_4 (0.2E_0)	19.2	17.2	24.1	17.8	16.6	19.6
T_5 (Control)	19.0	16.0	23.2	15.1	15.6	17.8
Average	20.72	18.38	24.64	18.6	17.3	21.6

Table 4.3.3 : Variation in leaf water potential (Mpa) at different irrigation levels (2003).

Treatment	Initial (12.3)	Wk 1 (20.3)	Wk 2 (26.3)	Wk 3 (1.4)	Wk 4 (7.4)	Wk 5 (17.4) MS*	Wk 5 (17.4) AS**
$T_1(0.8E_0)$	-0.75	-0.16	-0.10	-0.76	-0.96	-0.80	-1.72
$T_2(0.6E_0)$	-0.72	-0.17	-0.68	-0.82	-1.21	-0.82	-1.96
$T_3(0.4E_0)$	-0.78	-0.17	-0.88	-1.66	-1.30	-0.96	-2.08
$T_4(0.2E_0)$	-0.68	-0.18	-1.10	-1.86	-1.86	-1.48	-2.16
T_5(Control)	-0.82	-0.18	-1.26	-1.94	-1.92	-1.66	-2.20
Average	-0.75	-0.17	-0.80	-1.41	-1.45	-1.14	-2.02
SEm($\pm$)	-	0.083	0.320	0.266	0.162	0.165	0.036
C. D. at 5%	-	NS	0.187	0.818	0.499	0.508	0.111

Treatment	Wk 6 (22.4) MS	Wk 6 (22.4) AS	Wk 7 (29.4) MS	Wk 7 (29.4) AS	Wk 8 (6.5) MS	Wk 8 (6.5) AS	Wk 9 (13.5) MS	Wk 9 (13.5) AS
$T_1(0.8E_0)$	-0.60	-1.48	-0.65	-1.16	-0.94	-1.76	-0.98	-1.82
$T_2(0.6E_0)$	-0.74	-1.96	-0.64	-1.22	-1.02	-1.89	-1.15	-1.98
$T_3(0.4E_0)$	-0.89	-2.15	-0.68	-1.24	-1.66	-2.02	-1.69	-2.26
$T_4(0.2E_0)$	-1.08	-2.43	-0.69	-1.26	-1.84	-2.16	-2.23	-2.68
T_5(Control)	-1.71	-2.60	-0.83	-1.64	-2.11	-2.32	-2.40	-2.76
Average	-1.00	-2.12	-0.70	-1.30	-1.51	-2.03	-1.69	-2.30
SEm($\pm$)	0.061	0.084	0.237	0.103	0.056	0.045	0.133	0.120
C. D. at 5%	0.187	0.259	NS	0.318	0.173	0.139	0.411	0.369

*MS - Morning Sampling; **AS- Afternoon Sampling

Table 4.3.4 : Variation in leaf water potential (Mpa) at different irrigation levels (2004).

Treatment	Initial (5.3.04)	Wk 1 (18.3)	Wk 3 (1.4)	Wk 5 (15.4) Morning Sampling	Wk 5 (15.4) Afternoon Sampling
$T_1(0.8E_0)$	-0.68	-0.68	-0.84	-0.76	-1.23
$T_2(0.6E_0)$	-0.59	-0.72	-0.88	-0.82	-1.32
$T_3(0.4E_0)$	-0.61	-0.88	-1.36	-1.29	-1.83
$T_4(0.2E_0)$	-0.64	-0.94	-1.78	-1.61	-2.22
T_5(Control)	-0.67	-1.10	-1.82	-1.68	-2.36
Average	-0.64	-0.86	-1.34	-1.23	-1.79
SEm (±)	-	0.048	0.107	0.098	0.124
C. D. at 5%	-	0.148	0.329	0.302	0.381

Treatment	Wk 7 (29.4) Morning Sampling	Wk 7 (29.4) Afternoon Sampling	Wk 9 (13.5) Morning Sampling	Wk 9 (13.5) Afternoon Sampling
$T_1(0.8E_0)$	-0.74	-1.34	-0.78	-1.44
$T_2(0.6E_0)$	-0.84	-1.46	-0.92	-1.68
$T_3(0.4E_0)$	-1.31	-1.78	-1.41	-2.20
$T_4(0.2E_0)$	-1.72	-2.38	-1.78	-2.72
T_5(Control)	-1.81	-2.46	-1.84	-2.78
Average	-1.28	-1.88	-1.35	-2.16
SEm(±)	0.134	0.098	0.156	0.168
C. D. at 5%	0.414	0.310	0.482	0.519

Table 4.3.5 : Relative water content (RWC) (%) of leaf at different irrigation levels (2003).

Treatment	Initial (12.3)	Wk 1 (20.3)	Wk 2 (26.3)	Wk 3 (1.4)	Wk 4 (7.4)	Wk 5 (17.4)
$T_1(0.8E_0)$	94.82	95.15	93.78	92.66	91.50	91.45
$T_2(0.6E_0)$	94.50	94.98	93.16	91.99	90.32	90.64
$T_3(0.4E_0)$	95.02	94.15	91.51	90.01	88.00	87.89
$T_4(0.2E_0)$	95.10	93.82	90.81	87.70	86.03	86.11
T_5(Control)	94.76	93.78	89.22	87.31	84.87	83.75
Average	95.24	94.38	91.70	89.93	88.14	87.97
SEm(±)	-	8.601	0.463	0.647	0.564	0.873
C. D. at 5%	-	NS	1.425	1.994	1.737	2.688

Treatment	Wk 6 (22.4)	Wk 7 (29.4)	Wk 8 (6.5)	Wk 9 (13.5)	Average
$T_1(0.8E_0)$	91.04	92.15	93.45	92.28	92.61
$T_2(0.6E_0)$	89.41	91.81	92.17	90.87	91.71
$T_3(0.4E_0)$	86.57	89.16	90.22	88.77	89.59
$T_4(0.2E_0)$	84.38	84.92	86.34	85.89	87.33
T_5(Control)	82.11	84.23	84.58	84.35	86.02
Average	86.70	88.45	89.35	88.43	-
SEm(±)	0.610	1.288	0.969	0.572	-
C. D. at 5%	1.879	3.966	NS	1.763	-

Table 4.3.6 : Relative leaf water content (RWC) (%) at different irrigation levels (2004).

Treatment	Initial (5.3.04)	Wk 1 (18.3)	Wk 3 (1.4)	Wk 5 (15.4)	Wk 7 (29.4)	Wk 9 (13.5)	Average
$T_1(0.8E_0)$	95.00	93.55	91.65	92.13	92.10	91.20	92.60
$T_2(0.6E_0)$	94.89	92.69	90.24	90.96	91.18	90.25	91.70
$T_3(0.4E_0)$	95.38	90.66	86.94	87.57	86.69	85.71	88.82
$T_4(0.2E_0)$	95.13	87.92	84.35	84.91	85.11	83.19	86.77
$T_5(Control)$	94.91	87.06	83.21	84.08	84.26	82.15	85.94
Average	95.06	90.38	87.29	87.93	87.87	86.50	-
SEm(±)	-	0.832	0.971	1.070	1.385	1.280	-
C. D. at 5%	-	2.563	2.991	3.295	4.266	3.942	-

Table 4.3.7 (a) : Weekly increase in shoot growth at different irrigation levels (2003).

Treatment	Initial shoot length (cm) (11.3)	Percentage increase over previous week				
		Wk 1 (20.3)	Wk 2 (26.3)	Wk 3 (1.4)	Wk 4 (7.4)	Wk 5 (16.4)
$T_1(0.8E_0)$	10.52	3.44	3.42	3.26	2.46	2.75
$T_2(0.6E_0)$	13.02	3.55	6.75	2.85	2.18	2.18
$T_3(0.4E_0)$	11.52	3.51	3.08	2.19	1.39	4.67
$T_4(0.2E_0)$	11.54	4.14	2.19	2.75	3.00	1.77
$T_5(Control)$	11.13	5.18	3.32	1.97	2.39	0.80
Average	11.55	3.96	3.75	2.40	2.34	2.43
SEm(±)	-	0.242	0.397	0.327	0.261	0.319
C.D.at 5%	-	0.75	NS	NS	NS	NS

NS=Non-significant

Table 4.3.7 (b) : Weekly increase in shoot growth at different irrigation levels (2003).

Treatment	Initial shoot length (cm) (11.3)	Percentage increase over previous week				Total Increase (%)	Weekly average increase (%)
		Wk 6 (22.4)	Wk 7 (29.4)	Wk 8 (6.5)	Wk 9 (13.5)		
$T_1(0.8E_0)$	10.52	7.20	1.89	6.95	7.32	38.67	4.30
$T_2(0.6E_0)$	13.02	0.36	2.04	8.15	6.94	35.00	3.89
$T_3(0.4E_0)$	11.52	3.18	1.40	5.53	4.25	29.20	3.24
$T_4(0.2E_0)$	11.54	3.15	1.48	3.37	1.89	23.74	2.64
T_5(Control)	11.13	0.44	2.30	0.38	1.11	17.89	1.99
Average	11.55	2.87	1.82	5.68	5.10	28.90	3.21
SEm(±)	-	0.560	1.443	0.413	0.319	-	-
C.D.at 5%	-	NS	NS	1.273	0.984	-	-

Table 4.3.8 : Weekly increase in shoot growth at different irrigation levels (2004).

Treatment	Initial shoot length (cm) (5.3.04)	Percentage increase over previous week					Total Incre-ase (%)	Weekly average increase (%)
		Wk 1 (18.3)	Wk 3 (1.4)	Wk 5 (15.4)	Wk 7 (29.4)	Wk 9 (13.5)		
$T_1(0.8E_0)$	12.48	6.38	17.21	13.12	3.44	1.08	41.23	4.58
$T_2(0.6E_0)$	11.31	5.43	15.84	12.11	3.41	1.37	38.15	4.24
$T_3(0.4E_0)$	12.11	5.65	12.90	8.84	2.61	2.34	32.34	3.59
$T_4(0.2E_0)$	10.93	4.89	9.23	3.59	2.63	1.18	21.52	2.39
T_5(Control)	11.82	4.92	6.04	2.88	1.32	0.92	16.08	1.79
Average	11.73	5.45	12.32	8.11	2.68	1.38	-	-
SEm(±)	-	0.319	0.968	1.229	0.224	0.601	-	-
C.D.at 5%	-	0.984	2.985	3.787	0.698	NS	-	-

NS=Non-significant

Table 4.3.9 (a) : Net CO_2 assimilation and Stomatal conductance at different irrigation levels (2003)

Treatment	Week 1(20.3)		Week 2 (26.3)	
	Net CO_2 assimilation ($\mu molm^{-2}s^{-1}$)	Stomatal conductance ($milimolm^{-2}s^{-1}$)	Net CO_2 assimilation ($\mu molm^{-2}s^{-1}$)	Stomatal conductance ($milimolm^{-2}s^{-1}$)
T_1 (0.8E_0)	11.0	35.27	11.6	36.73
T_2 (0.6E_0)	10.5	28.70	10.5	22.36
T_3 (0.4E_0)	10.3	25.15	9.7	21.76
T_4 (0.2E_0)	9.8	23.30	9.5	20.16
T_5 (Control)	9.7	18.56	8.9	15.58
Average	10.28	26.20	9.96	23.32
SEm($\pm$)	0.329	2.094	0.298	1.109
C. D. at 5%	1.013	6.449	0.918	3.416

Treatment	Week 3 (1.4)		Week 4 (7.4)	
T_1 (0.8E_0)	9.6	32.94	7.8	24.83
T_2 (0.6E_0)	8.8	21.57	7.0	22.00
T_3 (0.4E_0)	7.8	19.90	6.6	16.30
T_4 (0.2E_0)	6.8	16.25	5.0	12.82
T_5 (Control)	5.4	14.16	4.5	9.70
Average	7.68	20.96	6.18	17.13
SEm($\pm$)	0.282	1.817	0.562	1.179
C. D. at 5%	0.868	5.597	1.732	3.631

Table 4.3.9 (b): Net CO$_2$ assimilation and Stomatal conductance at different irrigation levels (2003)

Treatment	Week 5 (16.4)		Week 6 (22.4)	
T$_1$ (0.8E$_0$)	7.2	27.10	7.0	28.40
T$_2$ (0.6E$_0$)	6.9	26.40	6.6	22.60
T$_3$ (0.4E$_0$)	5.8	22.63	5.4	21.56
T$_4$ (0.2E$_0$)	5.0	21.82	4.6	19.20
T$_5$ (Control)	4.8	15.60	4.0	17.70
Average	5.94	22.71	5.52	21.89
SEm($\pm$)	0.248	0.968	0.180	1.259
C. D. at 5%	0.763	2.981	0.554	3.877

Treatment	Week 7 (29.4)		Week 8 (6.5)	
T$_1$ (0.8E$_0$)	8.7	31.50	6.5	26.20
T$_2$ (0.6E$_0$)	7.2	29.65	6.1	24.71
T$_3$ (0.4E$_0$)	7.0	24.60	5.4	20.45
T$_4$ (0.2E$_0$)	6.6	18.91	5.1	16.60
T$_5$ (Control)	5.9	12.94	4.3	12.79
Average	7.08	23.52	5.46	20.15
SEm($\pm$)	0.393	1.487	0.233	1.164
C. D. at 5%	1.211	4.581	0.717	3.585

Table 4.3.10 : Net CO_2 assimilation and Stomatal conductance at different irrigation levels (2004).

Treatment	Initial (5.3)		Week 1 (18.3)	
	Net CO_2 assimilation (μmolm^{-2}s^{-1})	Stomatal conductance (milimolm^{-2}s^{-1})	Net CO_2 assimilation (μmolm^{-2}s^{-1})	Stomatal conductance (milimolm^{-2}s^{-1})
T_1 (0.8E_0)	9.8	29.86	12.8	38.40
T_2 (0.6E_0)	9.6	30.20	11.3	32.63
T_3 (0.4E_0)	9.5	28.66	9.6	30.11
T_4 (0.2E_0)	9.6	28.70	8.8	25.25
T_5 (Control)	9.7	29.50	7.5	20.50
Average	9.64	29.38	10.00	29.38
SEm ($\pm$)	-	-	0.395	1.376
C. D. at 5%	-	-	1.216	4.237

Treatment	Week 3 (1.4)		Week 5 (15.4)	
T_1 (0.8E_0)	10.8	28.67	8.4	27.66
T_2 (0.6E_0)	9.3	28.10	7.6	24.38
T_3 (0.4E_0)	8.1	26.18	7.1	23.66
T_4 (0.2E_0)	6.9	24.35	5.9	20.56
T_5 (Control)	6.2	23.56	5.2	18.80
Average	8.26	26.16	6.84	23.01
SEm ($\pm$)	0.432	0.529	0.253	1.245
C. D. at 5%	1.332	1.628	0.778	3.8368.2

Treatment	Week 7 (29.4)		Week 9 (13.5)	
T_1 (0.8E_0)	8.2	27.40	7.7	26.70
T_2 (0.6E_0)	7.4	25.28	6.3	24.44
T_3 (0.4E_0)	6.8	21.80	5.1	20.71
T_4 (0.2E_0)	5.3	20.10	4.2	19.10
T_5 (Control)	4.7	18.60	4.1	18.50
Average	6.48	22.64	5.48	21.89
SEm ($\pm$)	0.223	0.547	0.398	0.768
C. D. at 5%	0.688	1.685	1.225	2.366

Table 4.3.11(a) : Flowering and initial fruit set at different irrigation levels (2003).

Treatment	Length of panicle (cm)	Breadth of panicle (cm)	Total number of flowers /panicle
T_1 (0.8E_0)	22.59	12.82	1314.89
T_2 (0.6E_0)	21.87	13.33	1201.47
T_3 (0.4E_0)	23.58	12.61	1479.50
T_4 (0.2E_0)	23.91	13.87	1414.47
T_5 (Control)	24.54	14.31	1501.56
Average	23.30	13.39	1382.38
SEm (±)	1.216	0.803	100.022
C.D. at 5%	NS	NS	NS

NS=non-significant

Table 4.3.11(b) : Flowering and initial fruit set at different irrigation levels (2003).

Treatment	Number of staminate flowers /panicle	Number of pistillate flowers /panicle	Sex ratio (Staminate :Pistillate)	Initial fruits set /panicle
T_1 (0.8E_0)	901.34	413.55	2.53 : 1	14.65
T_2 (0.6E_0)	861.09	340.38	2.74 : 1	14.30
T_3 (0.4E_0)	1125.95	353.55	2.71 : 1	14.63
T_4 (0.2E_0)	1046.80	367.67	2.74 : 1	13.90
T_5 (Control)	1035.19	466.38	2.65 : 1	13.26
Average	994.07	388.31	2.67 : 1	14.15
SEm (±)	82.173	38.350	-	0.971
C.D. at 5%	NS	NS	-	NS

NS=non-significant

Table 4.3.12 (a) : Flowering and initial fruit set at different irrigation levels (2004).

Treatment	Length of panicle (cm)	Breadth of panicle (cm)	Total number of flowers /panicle
$T_1(0.8E_0)$	29.46	16.35	1806.71
$T_2(0.6E_0)$	28.38	15.92	1718.50
$T_3(0.4E_0)$	28.10	14.97	1621.65
$T_4(0.2E_0)$	27.75	15.15	1413.64
T_5(Control)	26.90	14.82	1281.36
Average	28.12	15.44	1566.57
SEm ($\pm$)	0.351	0.127	318.67
C.D. at 5%	1.081	0.391	981.50

NS=non-significant

Table 4.3.12 (b) : Flowering and initial fruit set at different irrigation levels (2004).

Treatment	Number of staminate flowers /panicle	Number of pistillate flowers /panicle	Sex ratio (Staminate :Pistillate)	Initial fruits set /panicle
$T_1(0.8E_0)$	1332.51	474.63	2.81:1	20.16
$T_2(0.6E_0)$	1269.83	448.39	2.83:1	19.58
$T_3(0.4E_0)$	1169.90	451.25	2.59:1	17.80
$T_4(0.2E_0)$	1060.76	352.37	3.01:1	16.25
T_5(Control)	943.12	338.91	2.79:1	15.10
Average	1155.22	413.12	2.81:1	17.78
SEm ($\pm$)	131.671	108.33	-	0.688
C.D. at 5%	NS	NS	-	2.118

NS=non-significant

151

Table 4.3.13 (a) : Weekly and total fruit drop at different irrigation levels (2003).

Treatment	Initial fruit set per panicle (26.3)	Percentage of fruit drop over previous week			
		1WAFS (1.4)	2 WAFS (7.4)	3 WAFS (16.4)	4 WAFS (22.4)
T_1 (0.8E_0)	14.65	6.15	13.02	2.61	0.45
T_2 (0.6E_0)	14.30	9.12	13.36	3.61	1.04
T_3 (0.4E_0)	14.63	14.26	17.54	3.52	4.77
T_4 (0.2E_0)	13.90	24.44	29.17	7.39	3.03
T_5 (Control)	13.26	16.07	33.64	6.67	6.56
Average	14.15	14.01	21.35	4.76	3.17
SEm (±)	0.971	0.386	2.553	0.612	0.589
C.D.at 5%	NS	NS	7.863	NS	NS

Table 4.3.13 (b) : Weekly and total fruit drop at different irrigation levels (2003).

Treatment	Initial fruit set per panicle (26.3)	Percentage of fruit drop over previous week			Fruits retained /panicle (at harvest)	Total drop (%)
		5 WAFS (29.4)	6 WAFS (6.5)	7 WAFS (13.5)		
T_1 (0.8E_0)	14.65	0.68	0.22	0.09	11.48	21.64
T_2 (0.6E_0)	14.30	1.44	0.13	0.21	10.55	26.22
T_3 (0.4E_0)	14.63	1.39	1.11	0.33	9.24	36.84
T_4 (0.2E_0)	13.90	2.27	1.57	0.96	6.37	54.17
T_5 (Control)	13.26	2.44	2.05	1.21	6.07	54.22
Average	14.15	1.64	1.02	0.56	8.78	38.62
SEm (±)	0.971	0.391	0.167	0.237	0.334	-
C.D.at 5%	NS	NS	0.514	NS	1.03	-

NS=non-significant

Table 4.3.14 (a) : Weekly and total fruit drop at different irrigation levels (2004).

Treatment	Initial fruit set per panicle (24.3)	Percentage of fruit drop over previous week			
		1WAFS (31.3)	2 WAFS (7.4)	3 WAFS (14.4)	4 WAFS (21.4)
T_1 (0.8E_0)	20.16	6.25	11.18	1.42	0.63
T_2 (0.6E_0)	19.58	9.41	12.16	2.25	0.91
T_3 (0.4E_0)	17.80	11.36	19.30	2.10	0.86
T_4 (0.2E_0)	16.25	14.41	23.46	5.43	3.02
T_5 (Control)	15.10	16.08	25.03	5.31	4.23
Average	17.78	11.50	18.23	3.30	1.93
SEm (±)	0.688	0.596	2.155	0.368	0.270
C.D.at 5%	2.118	1.837	6.639	1.135	0.831

NS=Non-significant

Table 4.3.14 (b) : Weekly and total fruit drop at different irrigation levels (2004).

Treatment	Initial fruit set per panicle (24.3)	Percentage of fruit drop over previous week			Fruits retained /panicle (at harvest)	Total drop (%)
		5 WAFS (28.4)	6 WAFS (5.5)	7 WAFS (12.5)		
T_1 (0.8E_0)	20.16	0.58	0.39	0.18	16.03	20.50
T_2 (0.6E_0)	19.58	0.60	0.12	0.21	14.72	24.80
T_3 (0.4E_0)	17.80	0.58	0.23	0.10	11.61	34.75
T_4 (0.2E_0)	16.25	0.98	0.63	0.31	8.41	48.25
T_5 (Control)	15.10	0.81	0.46	0.25	7.35	51.30
Average	17.78	0.71	0.37	0.21	11.62	35.92
SEm (±)	0.688	0.261	0.463	0.332	0.961	-
C.D.at 5%	2.118	NS	NS	NS	2.961	-

NS=Non-significant

Table 4.3.15 (a) : Weekly variations in peel, seed, aril and fresh weight of fruit due to different irrigation levels (2003).

Treatment	4 WAFS (22.4)				
	Fruit Wt. (g)	Peel Wt. (g)	Seed Wt. (g)	Aril Wt. (g)	Aril content (%)
T_1 (0.8E_0)	7.27	3.03	3.05	1.32	18.16
T_2 (0.6E_0)	7.06	2.95	3.03	1.17	16.55
T_3 (0.4E_0)	6.89	2.90	2.95	1.02	14.83
T_4 (0.2E_0)	6.88	2.80	2.91	1.00	14.41
T_5 (Control)	6.35	2.78	2.68	0.89	14.02
Average	6.86	2.83	2.92	1.11	16.18
SEm (±)	0.053	1.622	0.973	0.739	-
C.D. at 5%	0.163	NS	NS	NS	-

Treatment	5 WAFS (29.4)				
	Fruit Wt. (g)	Peel Wt. (g)	Seed Wt. (g)	Aril Wt. (g)	Aril content (%)
T_1 (0.8E_0)	12.87	3.82	4.37	4.86	37.80
T_2 (0.6E_0)	12.78	3.81	4.25	4.66	36.48
T_3 (0.4E_0)	12.75	3.72	4.35	4.58	35.92
T_4 (0.2E_0)	11.86	3.68	4.25	4.10	34.55
T_5 (Control)	11.80	3.46	3.73	4.08	34.58
Average	12.22	3.63	4.19	4.40	36.01
SEm (±)	0.076	0.825	0.933	0.038	-
C.D. at 5%	0.235	NS	NS	0.118	-

NS=Non-significant

Table 4.3.15 (b) : Weekly variations in peel, seed, aril and fresh weight of fruit due to different irrigation levels (2003).

Treatment	6 WAFS (6.5)				
	Fruit Wt. (g)	Peel Wt. (g)	Seed Wt. (g)	Aril Wt. (g)	Aril content (%)
T_1 (0.8E_0)	17.88	4.02	3.93	9.93	55.54
T_2 (0.6E_0)	17.83	4.01	4.13	9.69	54.35
T_3 (0.4E_0)	16.82	3.99	3.84	9.07	53.92
T_4 (0.2E_0)	16.66	3.91	3.99	8.76	52.59
T_5 (Control)	16.58	3.87	4.01	8.62	51.98
Average	17.15	3.96	3.98	9.21	53.68
SEm (±)	0.118	0.838	0.765	0.093	-
C.D. at 5%	0.365	NS	NS	0.287	-

Treatment	7 WAFS (13.5)				
	Fruit Wt. (g)	Peel Wt. (g)	Seed Wt. (g)	Aril Wt. (g)	Aril content (%)
T_1 (0.8E_0)	24.95	4.66	4.58	16.06	64.37
T_2 (0.6E_0)	24.85	4.61	4.49	15.81	63.62
T_3 (0.4E_0)	23.60	4.52	4.51	13.99	60.22
T_4 (0.2E_0)	22.13	4.41	4.43	12.66	58.72
T_5 (Control)	21.27	4.49	4.49	12.13	57.41
Average	23.14	4.54	4.48	14.13	60.87
SEm (±)	0.381	0.760	1.339	0.334	-
C.D. at 5%	1.174	NS	NS	1.030	-

NS=Non-significant

Table 4.3.16 (a) : Weekly variation in peel, seed, aril and fresh weight of fruits due to different irrigation levels (2004).

Treatment	4 WAFS (21.4)				
	Fruit Wt. (g)	Peel Wt. (g)	Seed Wt. (g)	Aril Wt. (g)	Aril content (%)
T_1 (0.8E_0)	7.02	2.99	2.80	1.23	17.52
T_2 (0.6E_0)	6.93	2.90	2.91	1.12	16.23
T_3 (0.4E_0)	6.81	2.88	2.88	1.05	15.50
T_4 (0.2E_0)	6.75	2.78	3.01	0.96	14.30
T_5 (Control)	6.25	2.75	2.63	0.87	13.95
Average	6.75	2.86	2.85	1.05	15.50
SEm (±)	0.039	0.877	1.011	0.681	-
C.D. at 5%	0.119	NS	NS	NS	-

Treatment	5 WAFS (28.4)				
	Fruit Wt. (g)	Peel Wt. (g)	Seed Wt. (g)	Aril Wt. (g)	Aril content (%)
T_1 (0.8E_0)	12.71	4.15	3.91	4.65	36.61
T_2 (0.6E_0)	12.69	4.21	4.12	4.36	34.33
T_3 (0.4E_0)	12.65	4.20	4.34	4.11	32.50
T_4 (0.2E_0)	11.78	4.15	3.97	3.66	31.10
T_5 (Control)	11.45	4.12	3.81	3.52	30.75
Average	12.26	4.17	4.03	4.06	33.06
SEm (±)	0.067	1.562	0.732	0.108	-
C.D. at 5%	0.208	NS	NS	0.334	-

NS=Non-significant

Table 4.3.16 (b) : Weekly variation in peel, seed, aril and fresh weight of fruits due to different irrigation levels (2004).

Treatment	6 WAFS (5.5)				
	Fruit Wt. (g)	Peel Wt. (g)	Seed Wt. (g)	Aril Wt. (g)	Aril content (%)
T_1 (0.8E_0)	18.31	4.21	3.98	10.12	55.62
T_2 (0.6E_0)	17.95	4.24	3.98	9.72	54.15
T_3 (0.4E_0)	16.78	4.22	3.55	9.01	53.70
T_4 (0.2E_0)	16.25	4.19	3.67	8.39	51.62
T_5 (Control)	16.10	4.15	3.68	8.27	51.35
Average	17.08	4.20	3.77	9.10	53.29
SEm (±)	0.179	1.380	1.734	0.286	-
C.D. at 5%	0.552	NS	NS	0.881	-

Treatment	7 WAFS (15.5)				
	Fruit Wt. (g)	Peel Wt. (g)	Seed Wt. (g)	Aril Wt. (g)	Aril content (%)
T_1 (0.8E_0)	24.71	4.48	4.22	16.01	65.21
T_2 (0.6E_0)	24.35	4.32	4.32	15.71	64.50
T_3 (0.4E_0)	23.10	4.50	4.35	14.25	61.25
T_4 (0.2E_0)	22.16	4.37	4.36	13.43	60.15
T_5 (Control)	22.23	4.42	4.28	13.53	60.86
Average	23.31	4.42	4.31	14.57	62.39
SEm (±)	0.400	1.337	2.605	0.398	-
C.D. at 5%	1.231	NS	NS	1.225	-

NS=Non-significant

Table 4.3.17 : Sun-burning and skin-cracking of fruit at different irrigation levels.

Treatment	5 WAFS		6 WAFS		7 WAFS		Normal fruits at harvest (%)
	Sun-burning (%)	Skin-cracking (%)	Sun-burning (%)	Skin-cracking (%)	Sun-burning (%)	Skin-cracking (%)	
T_1 (0.8E_0)	-	-	2.34 (1.68)*	1.81 (1.52)*	3.64 (2.03)*	2.20 (1.64)*	94.16 (9.70)*
T_2 (0.6E_0)	-	-	2.68 (1.78)	1.90 (1.55)	3.98 (2.12)	2.45 (1.72)	93.57 (9.67)
T_3 (0.4E_0)	1.62	-	4.87 (2.32)	3.45 (1.99)	7.81 (2.88)	3.71 (2.05)	88.48 (9.41)
T_4 (0.2E_0)	2.34	-	7.63 (2.85)	5.66 (2.48)	11.75 (3.43)	6.66 (2.68)	81.59 (9.03)
T_5 (Control)	3.35	-	9.20 (3.11)	7.53 (2.83)	12.95 (3.60)	9.67 (3.19)	77.38 (8.80)
Average	1.62	-	5.34	4.07	8.03	4.94	87.04
SEm (±)	-	-	0.510	0.140	0.542	0.578	1.529
C.D. at 5%	-	-	1.571	0.432	1.668	1.78	4.710

* Angular transformed values in parenthesis

Table 4.3.18 (a) : Effect of different irrigation levels on physical characters of fruit and yield (2003).

Treatment	Fruit length (cm)	Fruit diameter (cm)	Fresh fruit weight (g)	Peel weight (g)	Seed weight (g)
T_1 (0.8E_0)	4.09	3.44	24.95	4.57	4.37
T_2 (0.6E_0)	4.24	3.47	24.85	4.67	4.50
T_3 (0.4E_0)	4.13	3.45	23.60	4.66	4.70
T_4 (0.2E_0)	3.93	3.31	22.13	4.31	4.24
T_5 (Control)	4.2	3.39	21.27	4.63	4.33
Average	4.12	3.41	23.14	4.54	4.48
SEm (±)	0.101	0.893	0.381	0.311	0.299
C.D.at 5%	NS	NS	1.174	NS	NS

Table 4.3.18 (b) : Effect of different irrigation levels on physical characters of fruit and yield (2003).

Treatment	Pulp weight (g)	Aril content (%)	Number of fruits /tree	Yield (Kg/tree)	Impact on yield over dry treatment (T_5)
T_1 (0.8E_0)	16.06	64.37	2,711.48	67.65	+ 144.49 %
T_2 (0.6E_0)	15.81	63.62	2,630.91	65.38	+ 136.28 %
T_3 (0.4E_0)	13.99	60.22	2,069.23	48.85	+ 76.54 %
T_4 (0.2E_0)	12.66	58.72	1,657.32	31.38	+ 13.41 %
T_5 (Control)	13.13	57.41	1250.34	27.67	-
Average	14.13	60.87	2063.86	48.19	-
SEm (±)	0.523	-	137.497	5.114	-
C.D.at 5%	1.612	-	423.490	15.752	-

NS=Non-significant

Table 4.3.19 (a) : Effect of different irrigation levels on physical characters of fruit and yield (2004).

Treatment	Fruit length (cm)	Fruit diameter (cm)	Fresh fruit weight (g)	Peel weight (g)	Seed weight (g)
T_1 (0.8E_0)	4.25	3.45	24.71	4.48	4.22
T_2 (0.6E_0)	4.18	3.46	24.35	4.32	4.32
T_3 (0.4E_0)	4.11	3.43	23.10	4.50	4.35
T_4 (0.2E_0)	4.02	3.33	22.16	4.37	4.36
T_5 (Control)	4.09	3.38	22.23	4.42	4.28
Average	4.13	3.41	23.31	4.42	4.31
SEm (±)	1.381	0.962	0.400	1.337	2.605
C.D.at 5%	NS	NS	1.231	NS	NS

NS = Non-significant

Table 4.3.19 (a) : Effect of different irrigation levels on physical characters of fruit and yield (2004).

Treatment	Pulp weight (g)	Aril content (%)	Number of fruits /tree	Yield (Kg/tree)	Impact on yield over dry treatment (T_5)
T_1 (0.8E_0)	16.01	65.21	2981.60	71.25	118.56 %
T_2 (0.6E_0)	15.71	64.50	2806.10	68.10	108.90 %
T_3 (0.4E_0)	14.25	61.25	2215.65	49.75	52.61 %
T_4 (0.2E_0)	13.43	60.15	1695.31	37.20	14.11 %
T_5 (Control)	13.53	60.86	1488.46	32.60	-
Average	14.57	62.39	2237.42	51.78	-
SEm (±)	0.398	-	182.429	5.591	-
C.D.at 5%	1.225	-	561.880	17.221	-

NS = Non-significant

Table 4.3.20 : Influence of different irrigation levels on quality of fruit (2003).

Treatment	Total Soluble Solids (^{0}Brix)	Reducing sugar (% fresh wt.)	Total sugar (% fresh wt.)	Tritra- table acidity (%)	Sugar/a cid ratio	TSS/aci d ratio
T$_1$ (0.8E$_0$)	18.87	14.98	16.67	0.42	39.69	44.92
T$_2$ (0.6E$_0$)	18.60	15.12	17.01	0.48	35.44	38.75
T$_3$ (0.4E$_0$)	18.71	14.22	16.57	0.59	28.08	31.75
T$_4$ (0.2E$_0$)	18.10	14.15	17.33	0.57	30.40	31.74
T$_5$ (Control)	17.83	14.19	16.97	0.63	26.94	28.31
Average	18.42	14.53	16.91	0.54	32.11	35.09
SEm (±)	0.103	0.349	0.185	0.077	-	-
C.D.at 5%	0.317	NS	0.571	NS	-	-

NS = Non-significant

Table 4.3.21 : Influence of different irrigation levels on quality of fruit (2004).

Treatment	Total Soluble Solids (^{0}Brix)	Reducing sugar (% fresh wt.)	Total sugar (% fresh wt.)	Tritra- table acidity (%)	Sugar/a cid ratio	TSS/aci d ratio
T$_1$ (0.8E$_0$)	18.95	14.90	16.80	0.45	37.33	42.11
T$_2$ (0.6E$_0$)	18.80	15.21	17.10	0.47	36.38	40.00
T$_3$ (0.4E$_0$)	18.65	14.80	16.70	0.53	31.51	35.19
T$_4$ (0.2E$_0$)	18.30	14.96	17.02	0.49	34.73	37.35
T$_5$ (Control)	17.95	15.00	16.98	0.52	32.65	34.52
Average	18.53	14.97	16.92	0.49	34.52	37.83
SEm (±)	0.067	0.285	1.088	0.131	-	-
C.D.at 5%	0.205	NS	NS	NS	-	-

NS = Non-significant

Table 4.3.22 (a) : Water use (Et), Crop factor (Kc=Et/Ep) and Water Expense Efficiency (WEE) at different irrigation levels (2003).

Treatment	Week 1 (16.3.03-20.3.03)		Week 2 (21.3-26.3)		Week 3 (27.3-1.4)	
	Water use (Et) (mm)	Kc (=Et/Ep)	Water use (Et) (mm)	Kc (=Et/Ep)	Water use (Et) (mm)	Kc (=Et/Ep)
T_1 (0.8E_0)	36.20	0.858	10.56	0.480	29.00	0.806
T_2 (0.6E_0)	36.33	0.861	9.82	0.447	29.04	0.807
T_3 (0.4E_0)	36.92	0.875	8.76	0.398	29.78	0.827
T_4 (0.2E_0)	36.81	0.872	8.54	0.388	28.93	0.804
T_5 (Control)	36.50	0.865	7.50	0.341	29.70	0.825
Average	36.55	0.866	9.04	0.411	29.29	0.814

Treatment	Week 4 (2.4-7.4)		Week 5 (8.4-16.4)		Week 6 (17.4-22.4)	
	Water use (Et) (mm)	Kc (=Et/Ep)	Water use (Et) (mm)	Kc (=Et/Ep)	Water use (Et) (mm)	Kc (=Et/Ep)
T_1 (0.8E_0)	13.12	0.521	19.53	0.479	14.66	0.682
T_2 (0.6E_0)	11.75	0.466	17.22	0.422	13.72	0.638
T_3 (0.4E_0)	11.02	0.437	16.76	0.411	12.52	0.582
T_4 (0.2E_0)	10.19	0.404	16.44	0.403	12.22	0.568
T_5 (Control)	10.50	0.417	15.00	0.368	11.70	0.544
Average	11.32	0.449	16.99	0.416	12.96	0.603

Table 4.3.22 (b) : Water use (Et), Crop factor (Kc=Et/Ep) and Water Expense Efficiency (WEE) at different irrigation levels (2003).

Treatment	Week 7 (23.4-29.4)		Week 8 (30.4-6.5)		Week 9 (7.5-13.5)	
	Water use (Et) (mm)	Kc (=Et/Ep)	Water use (Et) (mm)	Kc (=Et/Ep)	Water use (Et) (mm)	Kc (=Et/Ep)
$T_1(0.8E_0)$	15.23	0.506	27.85	0.761	26.56	0.746
$T_2(0.6E_0)$	13.53	0.450	27.30	0.746	25.92	0.728
$T_3(0.4E_0)$	8.81	0.293	27.03	0.739	25.83	0.726
$T_4(0.2E_0)$	8.41	0.279	26.83	0.733	23.73	0.667
T_5(Control)	9.80	0.326	26.60	0.727	24.00	0.674
Average	11.16	0.370	27.12	0.741	25.21	0.708

Treatment	Total water use (6.3.03-6.5.03) (cm)	Water Expense Efficiency (WUE) (Kg/ha/cm)
$T_1 (0.8E_0)$	19.27	614.36
$T_2 (0.6E_0)$	18.46	619.80
$T_3 (0.4E_0)$	17.74	481.90
$T_4 (0.2E_0)$	17.21	319.09
T_5 (Control)	17.13	282.68
Average	17.96	463.57

Table 4.3.23 : **Water use (Et), Crop factor (Kc=Et/Ep) and Water Expense Efficiency (WEE) at different irrigation levels (2004).**

Treatment	Week 1 (9.3 - 15.3)		Week 3 (16.3 - 29.3)		Week 5 (30.3 – 15.4)	
	Water use (Et) (mm)	Kc (=Et/Ep)	Water use (Et) (mm)	Kc (=Et/Ep)	Water use (Et) (mm)	Kc (=Et/Ep)
T_1 (0.8E_0)	25.05	0.50	30.33	0.55	53.6	0.78
T_2 (0.6E_0)	23.10	0.46	26.15	0.47	51.7	0.75
T_3 (0.4E_0)	19.70	0.39	23.80	0.43	50.7	0.73
T_4 (0.2E_0)	17.50	0.35	21.50	0.39	47.2	0.68
T_5 (Control)	16.90	0.34	16.90	0.30	41.7	0.60
Average	20.45	0.41	23.74	0.43	48.98	0.71

Treatment	Week 7 (16.4 – 30.4)		Week 9 (1.5 - 15.5)		Total water use (cm) (9.3 – 15.5)	Water Expense Efficiency (WUE) (Kg/ha/cm)
	Water use (Et) (mm)	Kc (=Et/Ep)	Water use (Et) (mm)	Kc (=Et/Ep)		
T_1 (0.8E_0)	55.90	0.85	64.80	0.87	22.97	542.83
T_2 (0.6E_0)	52.80	0.81	58.90	0.79	21.26	560.56
T_3 (0.4E_0)	49.60	0.76	56.82	0.76	20.06	434.01
T_4 (0.2E_0)	49.40	0.75	55.11	0.74	19.07	341.37
T_5 (Control)	48.10	0.73	49.70	0.67	17.33	329.20
Average	51.16	0.78	57.06	0.76	20.14	441.59

Table 4.3.24 : Economics of drip irrigation in litchi (summary).

Fruit yield (t/ha) (average of four irrigation levels)	:	9.78
Gross return (Rs/ha)	:	2,44,500.00
Cost of cultivation (Rs/ha)	:	47,728.00
Net return (Rs/ha)	:	1,96,772.00
Benefit : Cost (average of four irrigation levels)	:	4.12 : 1
Benefit : Cost (irrigation at 80% pan coefficient, T_1)	:	5.47 : 1
Benefit : Cost (dry treatment, T_5)	:	1.81 : 1

Table 4.3.25 (a) :Computation of Economics of drip irrigation in litchi

A. Gross return
Yield of fruit (average of two years)
(Spacing : 7.5m x 7.5 m ≈ 178 trees/ha)

T1=(67.25+71.25)/2=69.45 kg/tree ≈12.36 t/ha
T2=(65.38+68.10)/2=66.74 kg/tree ≈ 11.88 t/ha
T3=(48.85+49.75)/2=49.30 kg/tree ≈ 8.77 t/ha
T4=(31.38+37.20)/2=34.29 kg/tree ≈ 6.10 t/ha
Average yield = 9.78 t/ha

Gross return:
Sale value of litchi fruit @ Rs. 25.00 per kg
Average of four irrigation levels = 9780 x 25 = Rs. 2,44,500.00
Irrigation at 80% pan coefficient (T_1) = 1236 x 25 =Rs. 3,09,000.00
Dry treatment (T_5) = 610 x 25 = Rs. 1,34,101.00

Table 4.3.25 (b) :Computation of Economics of drip irrigation in litchi

B. Cost of cultivation

1.	Cost of crop nutrition (manures & fertilizers)	Rs. 2,147.00

2.	Cost of plant protection
	   a) Chemicals (pesticides etc.)....Rs. 16,000.00
	   b) Hiring cost of net and generator for 10
	      days....Rs.2,500.00
		Rs. 18,500.00

3.	Soil cultivation	Rs. 3,600.00
4.	Labour cost (weeding, fertilization, spray, harvesting etc.)	Rs. 10,681.00
5.	Operational cost of drip irrigation system	Rs. 800.00
6.	10% of installation cost of drip irrigation	Rs. 12,000.00
	Total	*Rs. 47,728.00*

C. Net return and benefit : cost ratio

Net return:

Av. of four irrigation levels = 2,44,500.00–47,728.00 = Rs. 1,96,772.00

T_1 (0.8 E_0) = 3,09,000.00–47,728.00=Rs. 2,61,272.00

T_5 (Dry treatment) = 1,34,101.00–47,728.00 = Rs. 86,373.00

Benefit : cost ratio :

Average of four irrigation levels = 1,96,772.00 : 47,728.00 = 4.12 : 1

T_1 (0.8 E_0) = 2,61,272.00 : 47,728.00 =5.47 : 1

T_5 (Dry treatment) = 86,373.00 : 47,728.00 = 1.81 : 1

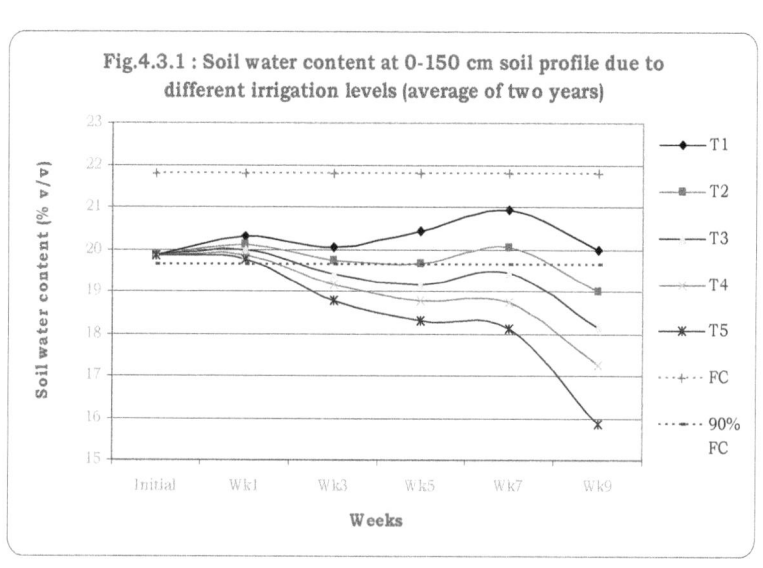

Fig.4.3.1 : Soil water content at 0-150 cm soil profile due to different irrigation levels (average of two years)

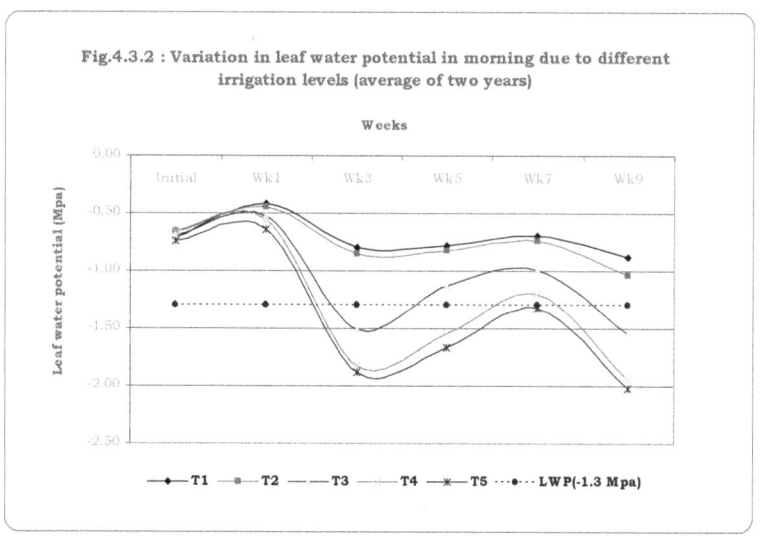

Fig.4.3.2 : Variation in leaf water potential in morning due to different irrigation levels (average of two years)

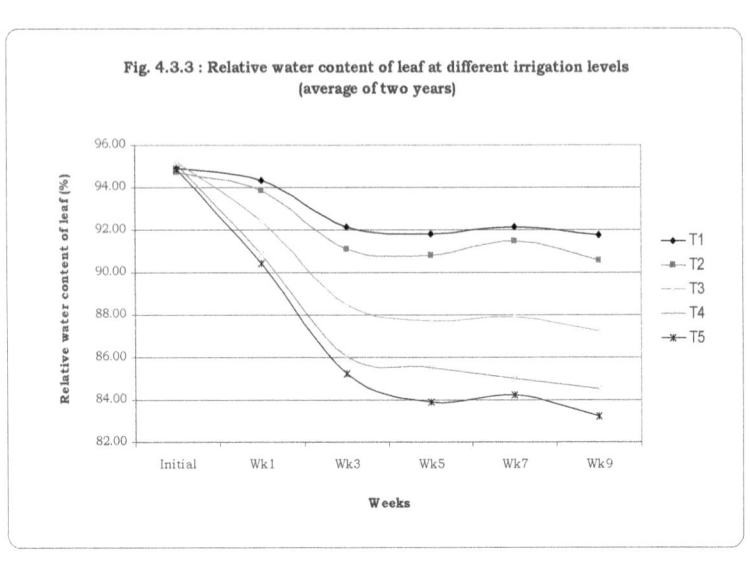

Fig. 4.3.3 : Relative water content of leaf at different irrigation levels (average of two years)

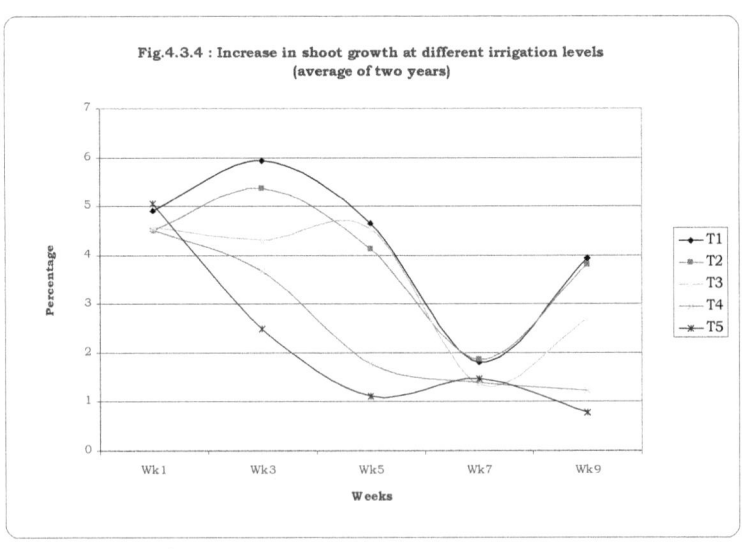

Fig.4.3.4 : Increase in shoot growth at different irrigation levels (average of two years)

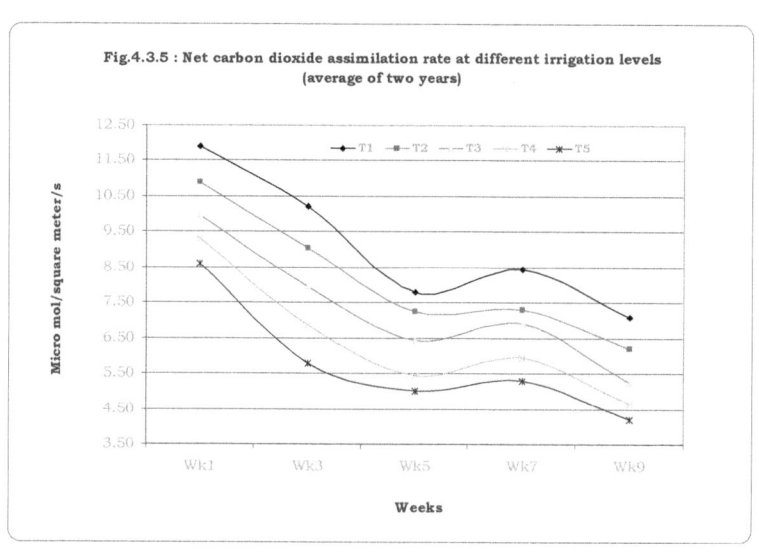

Fig.4.3.5 : Net carbon dioxide assimilation rate at different irrigation levels (average of two years)

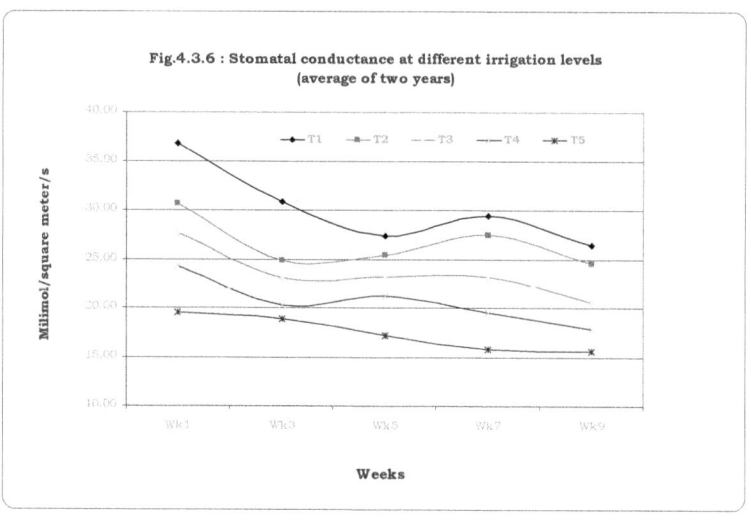

Fig.4.3.6 : Stomatal conductance at different irrigation levels (average of two years)

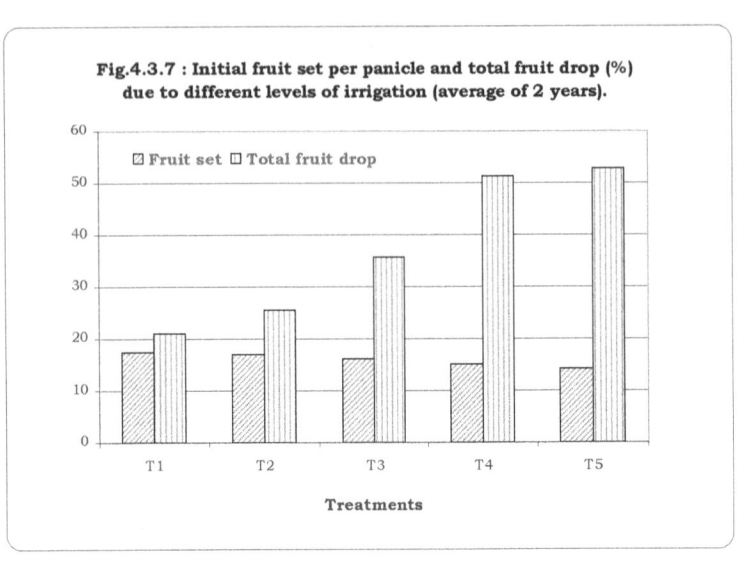

Fig.4.3.7 : Initial fruit set per panicle and total fruit drop (%) due to different levels of irrigation (average of 2 years).

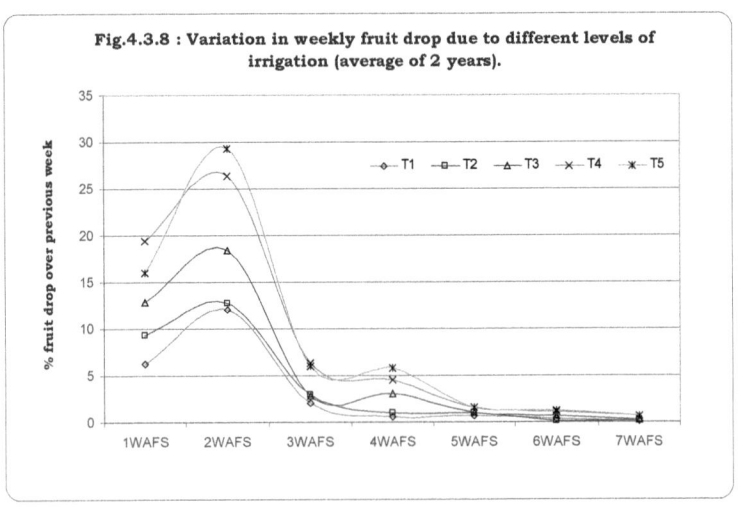

Fig.4.3.8 : Variation in weekly fruit drop due to different levels of irrigation (average of 2 years).

Fig. 4.3.9 : Sun-burning of fruit at different levels of irrigation.

Fig. 4.3.10 : Skin-cracking of fruit at different levels of irrigation.

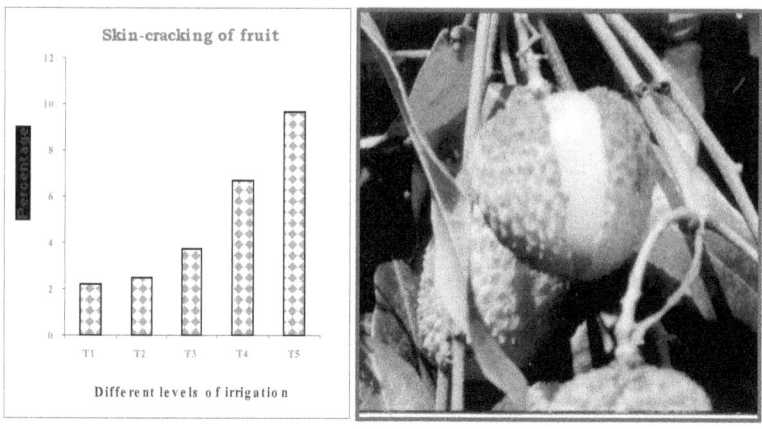

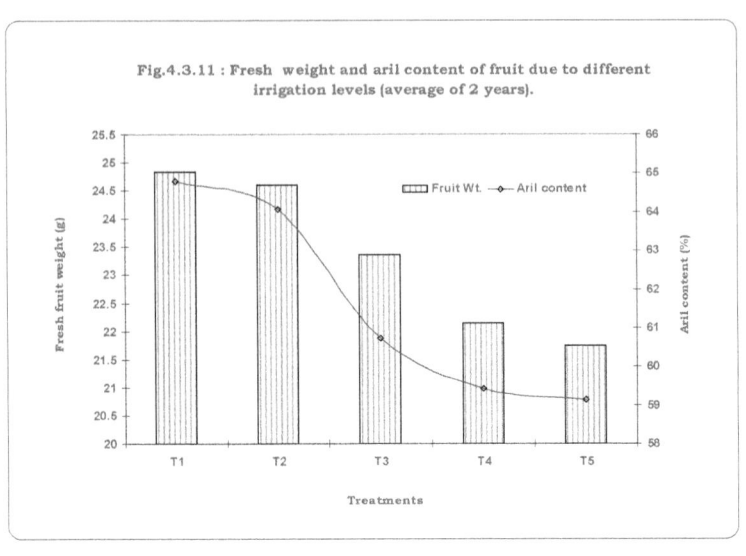

Fig.4.3.11 : Fresh weight and aril content of fruit due to different irrigation levels (average of 2 years).

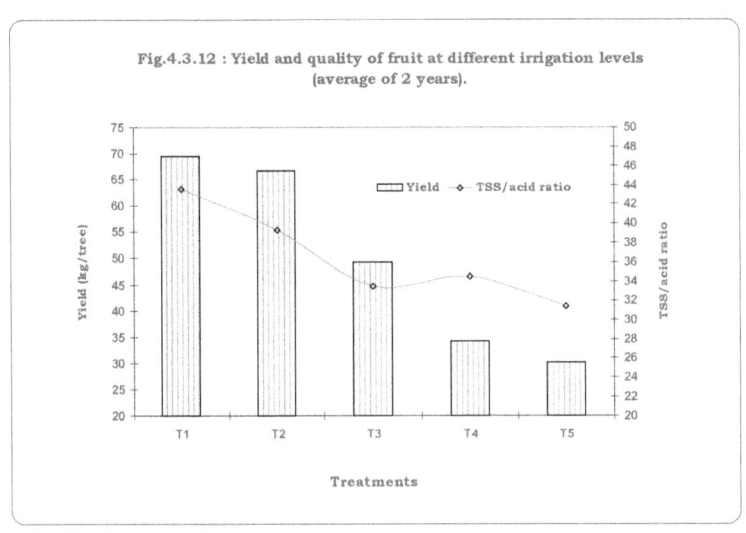

Fig.4.3.12 : Yield and quality of fruit at different irrigation levels (average of 2 years).

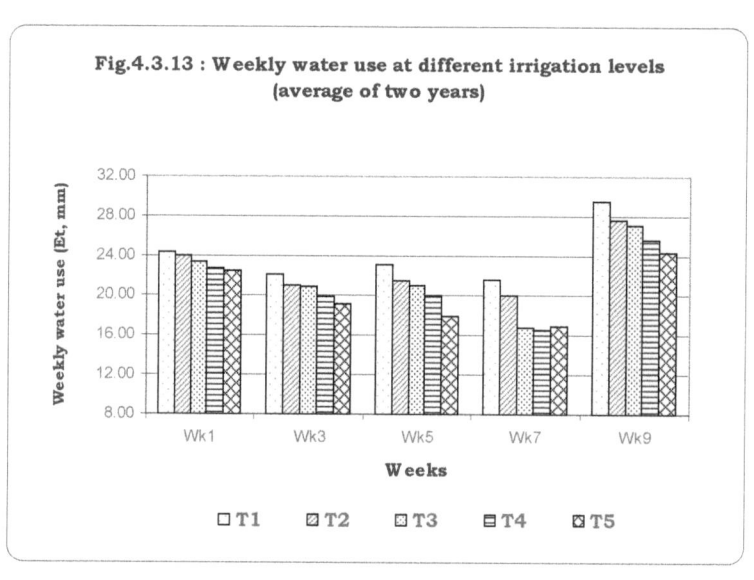

Fig.4.3.13 : Weekly water use at different irrigation levels (average of two years)

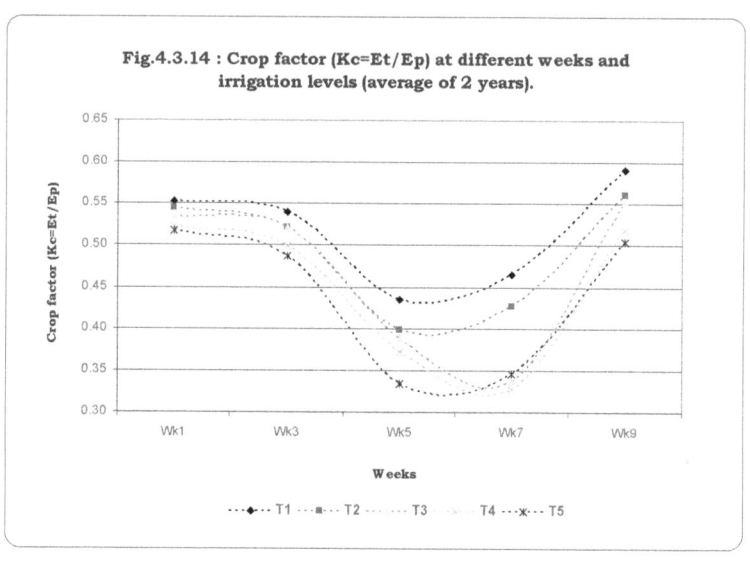

Fig.4.3.14 : Crop factor (Kc=Et/Ep) at different weeks and irrigation levels (average of 2 years).

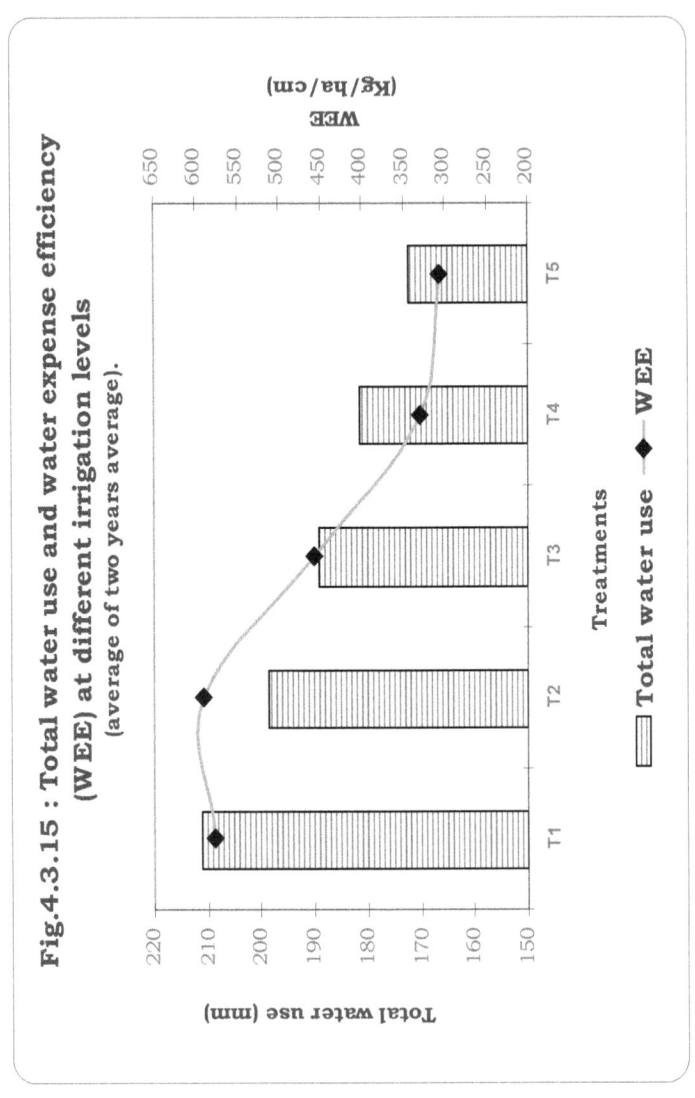

Fig.4.3.15 : Total water use and water expense efficiency (WEE) at different irrigation levels (average of two years average).

Plate 6 : Photographs showing the effect of different irrigation levels on fruit retention and yield.

5.0 SUMMARY AND CONCLUSION

Experiment I : Relationship between vegetative flushing and leaf N, timing of fertilizer N application.

In the present experiment, trees were fertilized with N, P_2O_5 and K_2O at 400, 200 and 400 g/tree/year, respectively. The P_2O_5 and K_2O were applied in two splits i.e., in July and March, while the time of N application constituted the treatments, viz., i) after harvest (July)(T_1), ii) in two splits, after harvest (July) and after fruit set (March) (T_2) iii) in two splits, after harvest (July) and in autumn (September) (T_3) iv) in three splits, June, September and March. (T_4) v) conventional method (as control) in two splits, onset and the end of monsoon (T_5).

The timing of N application showed marked variations in shoot growth and leaf N content in different months. Application of total N in July (T_1) or in two splits between June and September (T_3, T_5) caused higher (1.73%) leaf N content and more than 135% shoot growth between July and December, while application of one-third or half of total N during fruit development period (March) (T_4, T_2) showed lower (1.69%) leaf N content between fruit harvest and dormancy and reduced the post harvest shoot growth by about 11% to 18%. The increase in shoot growth during February to May, however, showed less than 4% variations due to different time of N application.

Maximum carbohydrate content in leaf (6.97%) and shoot (9.01%) in the month of December and maximum number of flowering shoots (40.60%) were recorded when N was applied in two splits in July and March (T_2), while application of N in July and September (T_3) showed minimum leaf carbohydrate content (6.17%) and flowering (29.78%).

Application N after fruit set (T_2, T_4) showed higher retention (13.30 to 13.72)of fruit per panicle compared with 11.80 to 12.83 fruits per panicle when N was not applied after fruit set (T_1, T_3, T_5). Trees receiving N at 400g/tree in two splits in July and March (T_2) produced maximum number of fruits (2032.24/tree) and highest yield (43.83 kg/tree) compared with

1768.06 fruits and 35.82 kg yield in trees receiving N at 400 g/tree in June and August (T_3).

Maximum fruit weight (23.43 g) and aril content (59.81%) were recorded when 50% N was applied after fruit set (T_2). Significantly higher total sugars content (15.71 to 15.74%) of fruit was recorded due to application of fertilizer N after fruit set (T_2, T_4). Highest TSS content and lowest tritratable acidity (0.48%) content of fruit were recorded due to T_2 treatment. The TSS/acid ratio of fruit was more than 38 due to T_2, T_4 and T_5 and less than 35 due to T_1 and T_3 treatments.

The correlation studies revealed that the pre-flowering (October to December) leaf N content had a positive relation with shoot growth rate during October to December but showed a negative relation with carbohydrate content of leaf and shoot in December, number of flowering shoot (%) and yield. The pre-flowering carbohydrate content of leaf and shoot (in December) showed significant and positive correlation with the number of flowering shoot (%) and yield. The leaf N content during fruit growth and development period (April to May) showed significant and positive correlation with fruit retention and also positive relationships with yield and quality (TSS/acid ratio) of fruit. The carbohydrate content of both leaf and shoot in March had also showed positive correlation with retention, yield and quality of fruit but it was the shoot carbohydrate content in March that showed significant and positive relation with fruit yield.

It may be suggested that the nitrogenous fertilizer should be applied in two splits in July (after fruit harvest) and March (after fruit set) which caused 17.59% increase in yield and improved fruit quality over the conventional timing of fertilizer application (in June and August) in West Bengal.

Experiment II : Refinement of leaf and soil sampling

In this experiment the variables of leaf and soil sampling were studied for standardization of leaf and soil sampling techniques and development of a leaf nutrient standard.

Considerable variations in leaf nutrients composition and soil nutrients content were recorded due to the variables of leaf and soil sampling. The leaf nutrient (N, P and K) content, in general, were lower during vegetative dormancy (January) compared with fruit growth and development period (March to May) and post harvest vegetative growth period (July to November).

The N, P and K content of leaf was higher in the east direction compared with west, north and south directions of tree canopy. The second pair of leaflets of the second leaf from tip of the shoot showed maximum content of leaf N (1.70%-1.72%), P (0.20%-0.21%) and K (1.10%-1.18%) which were minimum (1.50%-1.65%, 0.14%-0.18% and 0.95%-0.98%, respectively) in the sixth leaf (2nd pair). However, the first, third and fourth leaves showed very little variations in N, P and K content. The nutrients (N, P and K) content of leaf within 1 m of canopy height were low (1.30%-1.61% N, 0.12%-0.16% P and 0.72%-1.10% K, respectively) which gradually increased with sampling height up to 3 m (1.38%-1.71% N, 0.15%-0.20% P and 0.77%-1.18% K, respectively) and then declined marginally up to the sampling height of 5 m or above (1.51%-1.68% N, 0.14%-0.19% P and 0.73%-1.14% K, respectively).

The available N, P and K content of soil, in general, showed maximum depletion between January and May (flowering and fruit development period) which then increased in the month of July following fertilization after fruit harvest. It declined again in September and then increased between November and January. The organic carbon content of soil, however, was minimum (0.536%) in winter (January) compared with 0.627% to 0.632% in monsoon (July to September).

The top soil (0-30 cm) layer showed highest concentration of organic carbon (0.621%), available N (285.80 kg/ha), P (54.00 kg/ha) and K (136.90 kg/ha) which declined with increase in depth of sampling and it was recorded 0.583% organic carbon, 278.90 kg N/ha, 50.89 kg P/ha and 132.80 kg K/ha at 90 cm soil depth.

Considering temporal and spatial variations in soil nutrients content, it appears that a composite sample should be drawn for estimation taking soil samples randomly from atleast 10 spots (per hectare) at 0-90 cm soil depth. The sample may be drawn in the month of March for estimating soil nutrient status and fertilizer recommendation.

Considering the variations in leaf nutrients content due to the variables of leaf sampling, it may be suggested that a representative, composite sample of 16-20 pair of leaflets should be collected from four directions (east, west, north and south of atleast four trees per hectare) at the canopy height between 2 m and 4 m from ground level, taking the second pair of leaflets of the second leaf from tip of the shoot. At anthesis, the mature leaf (2nd pair) just behind the panicle should be sampled similarly as the suitable 'index tissue' for estimating nutrition status of bearing litchi tree cv. Bombai.

The pre-flowering (December) leaf nutrient status may be estimated for prediction of flower bud differentiation and flowering intensity. If the leaf N content ranged between 1.40% and 1.50% with leaf and shoot C/N ratio ranging between 4.5-5.5 and 7.5-8.5, respectively, atleast 40% of shoots on a tree may be expected to bear panicle in the spring.

The tree and soil nutrition status should be estimated at anthesis (March) for optimum yield and quality of fruit. The available soil N content of nearly 300.0 kg/ha and 1.65% to 1.75% leaf N content at anthesis (March) support a yield of 43.83 kg fruit/tree when 200 g N/tree was applied after fruit set (March) and other factors (irrigation, climate, pest and diseases) were not yield limiting. If the leaf N content at anthesis ranged between 1.65% and 1.75%, total N (400 g/tree) should be applied in two splits in July (after fruit harvest) and March (after fruit set) for optimum yield and better quality of fruit in 22 years old 'Bombai' litchi trees grown in alluvial region of West Bengal. Based on the adequate leaf nutrient ranges of N : 1.65-1.75%, P : 0.18-0.20% and K : 1.00-1.20% and soil nutrients content of about 302 kg available N/ha, 62 kg available K_2O/ha and 140 kg available K_2O/ha at anthesis, we suggest to apply N, P_2O_5 and K_2O at 400, 200 and 400 g/tree/year in two splits in July (after fruit harvest) and March (after fruit set) for a sustainable yield target of about 8.0 t/ha from > 20 years old 'Bombai' litchi orchard grown in alluvial region of West Bengal.

Experiment III : To study the effects of water relations and CO_2 assimilation on growth and yield of field grown trees.

In this experiment, soil moisture was supplemented from anthesis through drip irrigation system at 80, 60, 40, 20 and 0% of pan coefficient (i.e., five treatments). The effect of different levels of irrigation on soil water content, shoot growth, CO_2 assimilation rate, yield and quality of fruit have been summarized below.

Application of irrigation at 80% pan coefficient (T_1) showed the highest soil water content (SWC) (20.00% to 20.93%) while lowest SWC (15.87% to 19.77%) was recorded due to dry treatment (T_5). In other words, there was the minimum fluctuations in SWC which was almost static at 93.30% of field capacity (F.C.) in all the weeks by irrigating at 80% pan coefficient (T_1) compared with the maximum fluctuations in SWC (90.7% of F.C. in week 1 to 72.8% of field capacity in week 9) due to dry treatment (T_5). At lower irrigation levels, the SWC fell below 90% of field capacity at week 3 (T_3, T_4 and T_5), week 5 (T_3, T_4 and T_5), week 7 (T_3, T_4 and T_5) and week 9 (T_2, T_3, T_4 and T_5).

The morning leaf water potential (LWP) was above -1.30 Mpa in all the weeks by irrigating at 60% and 80% pan coefficient (T_2 and T_1). At lower irrigation levels, the morning LWP fell below -1.30 Mpa at week 3 (T_3, T_4 and T_5), week 5 (T_4 and T_5), week 7 (T_5) and week 9 (T_3, T_4 and T_5). Irrigation at 80% pan coefficient (T_1) caused higher LWP in the afternoon (-1.44 Mpa to -1.23 Mpa) compared with afternoon LWP of -2.78 Mpa to -2.36 Mpa in dry treatment (T_5). At lower irrigation levels, the afternoon LWP fell below -2.0 Mpa at week 5 (T_4 and T_5), week 7 (T_4 and T_5) and week 9 (T_3, T_4 and T_5). The highest relative water content (RWC) of leaf (91.74 to 94.35%) was recorded due to irrigation at 80% pan coefficient (T_1) compared with the lowest (83.25% to 90.42%) in dry treatment (T_5). The decline in RWC of leaf below 90% corresponded to the LWP below -1.30

Mpa under when the levels of applied irrigation decreased at different weeks.

Irrigation at 80% pan coefficient (T_1) caused the highest (39.95%) increase in total shoot growth compared with the lowest (16.98%) in dry treatment (T_5). The increase in growth due to irrigation levels, however, showed higher variations at week 9 (0.78 to 3.93%), week 5 (1.12 to 4.65%) and week 3 (2.49 to 5.93%) compared with week 1 (4.49 to 5.05%) and week 7 (1.40 to 1.81%).

The levels of irrigation (treatments) caused marked variations in stomatal conductance and net CO_2 assimilation rate in all the weeks. Irrigation at 80% pan coefficient (T_1) caused the highest stomatal conductance (26.45 to 36.84 milimolm^{-2}s^{-1}) and net CO_2 assimilation rate (7.10 to 11.90 µmolm^{-2}s^{-1}) compared with the lowest stomatal conductance (15.65 to 19.53 milimolm^{-2}s^{-1}) and net CO_2 assimilation rate (4.20 to 8.60 µmolm^{-2}s^{-1}) in dry treatment (T_5). In week 3 (post fruit set stage), the stomatal conductance fell below 25.00 milimolm^{-2}s^{-1} and the net CO_2 assimilation rate fell below 9.00 µmolm^{-2}s^{-1} due to low levels of irrigation (T_3, T_4 and T_5) treatments that corresponded with the SWC (0-150 cm depth) below 90% of field capacity and morning LWP below -1.30 Mpa. The decline in stomatal conductance and net CO_2 assimilation rate as the leaf water potential fell (below -1.30 Mpa) under dry (T_5) and partial dry (T_4, T_3) treatments was pronounced in all the weeks.

The flowering characters showed significant variations in 2004 but were non-significant in 2003. Maximum length (29.46 cm) and breadth (16.35 cm) of panicle in 2004 were recorded by irrigating at 80% pan coefficient, while it was minimum (26.90 cm and 14.82 cm, respectively) in dry treatment. The number of staminate and pistillate flowers per panicle showed no specific trend with different irrigation levels in 2003, but the number of both types of flowers per panicle in 2004 showed gradual decline with decrease in the levels of irrigation. The ratio of the number of staminate and pistillate flowers on a panicle varied between 2.53 : 1 and 2.74 : 1 in 2003 and between 2.59 : 1 and 3.01 : 1 in 2004, but showed no specific trend with irrigation levels. The number of fruits set per panicle showed gradual decrease at the lower levels of irrigation treatments in both the years. Irrigating at 80% pan coefficient caused maximum fruit set (14.65/panicle in 2003 and 20.16/panicle in 2004) compared with the lowest (13.26/panicle in 2003 and 15.01/panicle in 2004) in dry treatment.

The period between 1st and 3rd week after fruit set (WAFS) was found very sensitive to moisture stress when lower levels of irrigation (at or below 40% pan coefficient) as well as dry treatment caused more than 12.5% fruit drop at 1 WAFS and more than 18.0% fruit drop at 2 WAFS. The total drop of fruit (until maturity) was lowest (21.07%) at the highest level of irrigation $(T_1:0.8\ E_0)$ treatment, while the dry treatment caused the maximum fruit drop of 52.76 percent. The decline in net CO_2 assimilation rate as LWP fell below -1.30 Mpa under dry (T_5) and partial dry (T_4, T_3) treatments appeared to have adverse effect on retention and growth and development of fruit.

Moisture stress showed significant influence on the intensity of sun-burning and skin-cracking of fruit. The minimum intensity of sun-burning and skin-cracking were recorded by irrigating at 80% pan coefficient (3.64% and 2.20%, respectively), while it was as high as 12.95% and 9.67%, respectively in dry treatment (T_5). More than 93% of the harvested fruits were normal by application of irrigation during fruit growth and development period at 60% and 80% pan coefficient $(T_2$ and $T_1)$.

Application of irrigation at 80% and 60% pan coefficient $(T_1$ and $T_2)$ significantly increased the weight of fruit and aril content over the dry treatment (T_5) in both the years. The weight of fruit was minimum (21.27g to 22.23g) in both the years due to dry treatment (T_5) compared with 24.71 to 24.95g in T_1 $(0.8\ E_0)$ treatment. Moisture stress due to application of irrigation at or below 40% pan coefficient $(T_3, T_4$ and $T_5)$ reduced the aril content (< 61.5%) of fruit which was significantly higher (> 63.5%) at irrigation levels of 60% and 80% pan coefficient $(T_2$ and $T_1)$.

More than 2500 fruits per tree were produced on the trees that received irrigation at 60% and 80% pan coefficient however, the yield markedly reduced to less than 2250 fruits per tree at lower levels (at or below 40%) of irrigation treatments. The highest yield of 67.65 kg fruit per tree in 2003 and 71.25 kg fruit per tree in 2004 were recorded by irrigating at 80% pan coefficient (T_1) compared with 27.67 kg fruit per tree in 2003 and 32.60 kg fruit per tree in 2004 in dry treatment (T_5).

Different levels of irrigation showed significant effect on total soluble solids (TSS) content of fruit in both the years. The TSS content of fruit was recorded highest (18.87^0B in 2003 and 18.95^0B in 2004) by irrigating at 80% pan coefficient (T_1) compared with the lowest (17.83^0B in 2003 and 17.95^0B in 2004) in dry treatment (T_5). Maximum content of reducing sugar was recorded (15.12% in 2003 and 15.21% in 2004) by irrigating at 60% pan coefficient (T_2). The highest amount total sugar content of fruit was recorded (17.33%) in 2003 due to T_4 (0.2 E_0) and 17.10% in 2004 due to T_2 (0.6 E_0) treatments. Irrigating at 80% pan coefficient caused the minimum (0.42% in 2003 and 0.45% in 2004) fruit acidity which was maximum in T_5 (0.63%) in 2003 and T_3 (0.53%) treatments in 2004. An overall decrease in the TSS/acid ratio of fruit was found with the increasing intensity of moisture stress. The maximum TSS/acid ratio of 44.92 in 2003 and 42.11 in 2004 were recorded by irrigating at 80% pan coefficient.

Different irrigation levels caused marked variations in weekly water use (Et) which varied between 24.14±3.95 mm in highest level of irrigation (T_1:0.8 E_0) and 20.17±3.75 mm in dry treatment (T_5). The crop factor (Kc) of litchi cv. Bombai (average of two years) varied between 0.33 and 0.59 during the fruit growth and development period. At higher irrigation level of 80% pan coefficient (T_1), the Kc was higher (0.43 to 0.59) compared with 0.33 to 0.52 in dry treatment (T_5). The overall value of Kc between anthesis and harvest was 0.52 by irrigating at 80% pan coefficient, which declined at lower levels (T_2, T_3 and T_4) of irrigation and was found lowest (0.44) in dry treatment (T_5). The total water use (Et) of 15 years old Bombai litchi trees between anthesis and harvest was 190.50 mm with a corresponding water expense efficiency (WEE) of 452.50 kg/ha/cm. The total Et was recorded maximum (211.20 mm) by irrigating at 80% pan coefficient (T_1) and minimum (172.30 mm) in dry treatment (T_5). However, the WEE was maximum (590.18 kg/ha/cm) when irrigated at 60% pan coefficient (T_2), closely followed by 578.18 kg/ha/cm in T_1 (0.8 E_0) treatment. There was a sharp decline in WEE at lower irrigation levels (at or below 40% pan coefficient, T_3 and T_4) and it was lowest (305.94 kg/ha/cm) in dry treatment (T_5).

The benefit : cost ratio of litchi cultivation under drip irrigation (average of four irrigation levels) was 4.12 : 1. Irrigation at 80% pan coefficient (T_1) caused net return of Rs. 2,61,272.00 per hectare with a benefit : cost ratio of 5.47 : 1, while the net return was very low (Rs. 1,34,101.00) in dry treatment (T_5) resulting in a very poor benefit : cost ratio of 1.81 : 1.

From the present study it appears that adequate soil moisture during fruit growth and development period (March to May) is essential to minimize fruit drop and for better development of fruit, higher yield and improved quality of fruit. During this period, the leaf water potential in the morning should not fall below -1.30 Mpa to ensure maximum rate of stomatal conductance and net CO_2 assimilation that influence the fruit development and yield. It can be achieved when soil water content is maintained at near field capacity by irrigating at 80% pan coefficient which is equivalent to an overall crop factor of 0.52 (for irrigation scheduling) and application of irrigation to 15 years old bearing litchi trees cv. Bombai grown on sandy loam soil from anthesis to harvest (March to May) should be by drip irrigation using 2 drippers tree^{-1} discharging 8 litre water hour^{-1} dripper^{-1} for 3 hours at alternate day. The corresponding water expense efficiency and benefit : cost ratio estimated as 578.18 kg/ha/cm and 5.47 : 1, respectively.

6.0 SYNOPSIS

Considering the commercial importance of litchi in West Bengal and the research needs in litchi production technologies, the following experiments were conducted for *"Optimizing nutrition programme and irrigation for sustainable litchi production"* in West Bengal with the following objectives :

Exp.I : Timing of fertilizer N application : To develop relationship between vegetative flushing and leaf N, timing of fertilizer application at different stages.

Exp. II : Refinement of leaf and soil sampling : To develop leaf and soil sampling techniques, leaf nutrient standards.

Exp. III : Effect of water relations and CO_2 assimilation : To study the effects of water relations, CO_2 assimilation on growth and yield of field grown trees.

The experiments were conducted at the Horticultural Research Station of Bidhan Chandra Krishi Viswavidyalaya, Mondouri, Nadia, West Bengal during 2001 to 2004 on 22 years old (Exp.I & II) and 15 years old (Exp. III) litchi trees. The materials and methods and results of these experiments have been summarized below.

Experiment I : Relationship between vegetative flushing and leaf N, timing of fertilizer N application.

In this experiment, trees were fertilized with N, P_2O_5 and K_2O at 400, 200 and 400 g/tree/year, respectively. The P_2O_5 and K_2O were applied in two splits i.e., in July and March, while the time of N application constituted the treatments, viz., i) after harvest (July)(T_1), ii) in two splits, after harvest (July) and after fruit set (March) (T_2) iii) in two splits, after harvest (July) and in autumn (September) (T_3) iv) in three splits, June, September and March. (T_4) v) conventional method (as control) in two splits, onset and the end of monsoon (T_5). Observations were recorded on i) monthly increase in shoots growth, ii) details of flowering, yield and fruit quality, iii) monthly N content of leaf and annually after fruit set for P, K, Ca, Mg, Fe, Zn, Cu and

B, iv) carbohydrate reserves in leaf and shoots at pre-flowering stage (December) and at anthesis (March), v) The relation between pre-flowering (December) plant nutrient status and flowering, vi) The relation between plant nutrient status at anthesis (March) and yield and quality of fruit.

The timing of N application showed marked variations in shoot growth and leaf N content in different months. Application of total N in July (T_1) or in two splits between June and September (T_3, T_5) caused higher (1.73%) leaf N content and more than 135% shoot growth between July and December, while application of one-third or half of total N during fruit development period (March) (T_4, T_2) showed lower (1.69%) leaf N content between fruit harvest and dormancy and reduced the post harvest shoot growth by about 11% to 18%. The increase in shoot growth during February to May, however, showed less than 4% variations due to different time of N application.

Maximum carbohydrate content in leaf (6.97%) and shoot (9.01%) in the month of December and maximum number of flowering shoots (40.60%) were recorded when N was applied in two splits in July and March (T_2), while application of N in July and September (T_3) showed minimum leaf carbohydrate content (6.17%) and flowering (29.78%).

Application of N after fruit set (T_2, T_4) showed higher retention (13.30 to 13.72) of fruit per panicle compared with 11.80 to 12.83 fruits per panicle when N was not applied after fruit set (T_1, T_3, T_5). Trees receiving N at 400g/tree in two splits in July and March (T_3) produced maximum number of fruits (2032.24/tree) and highest yield (43.83 kg/tree) compared with 1768.06 fruits and 35.82 kg yield in trees receiving N at 400 g/tree in June and August (T_3).

Maximum fruit weight (23.43 g) and aril content (59.81%) were recorded when 50% N was applied after fruit set (T_2). Significantly higher total sugars content (15.71 to 15.74%) of fruit was recorded due to application of fertilizer N after fruit set (T_2, T_4). Highest TSS content and lowest tritratable acidity (0.48%) content of fruit were recorded due to T_2

treatment The TSS/acid ratio of fruit was more than 38 due to T_2, T_4 and T_5 and less than 35 due to T_1 and T_3 treatments.

The correlation studies revealed that the pre-flowering (October to December) leaf N content had a positive relation with shoot growth rate during October to December but showed a negative relation with carbohydrate content of leaf and shoot in December, number of flowering shoot (%) and yield. The pre-flowering carbohydrate content of leaf and shoot (in December) showed significant and positive correlation with the number of flowering shoot (%) and yield. The leaf N content during fruit growth and development period (April to May) showed significant and positive correlation with fruit retention and also positive relationships with yield and quality (TSS/acid ratio) of fruit. The carbohydrate content of both leaf and shoot in March had also showed positive correlation with retention, yield and quality of fruit but it was the shoot carbohydrate content in March that showed significant and positive relation with fruit yield.

It may be suggested that the nitrogenous fertilizer should be applied in two splits in July (after fruit harvest) and March (after fruit set) which caused 17.59% increase in yield and improved fruit quality over the conventional timing of fertilizer application (in June and August) in West Bengal.

Experiment II : Refinement of leaf and soil sampling techniques

Based on yield performance of the trees of experiment-I, the trees under treatment 2 (highest yield in 2001) were selected for experiment-II. In this experiment the variables of leaf (directions, age and height of sampling) and soil (temporal and spatial) sampling were studied for standardization of leaf and soil sampling techniques and development of a leaf nutrient standard.

Considerable variations in leaf nutrients composition and soil nutrients content were recorded due to the variables of leaf and soil sampling. The leaf nutrient (N, P and K) content, in general, were lower during vegetative dormancy (January) compared with fruit growth and development period (March to May) and post harvest vegetative growth period (July to November).

The N, P and K content of leaf was higher in the east direction compared with west, north and south directions of tree canopy. The second pair of leaflets of second leaf from the shoot tip showed maximum content of leaf N (1.70%-1.72%), P (0.20%-0.21%) and K (1.10%-1.18%) which were minimum (1.50%-1.65%, 0.14%-0.18% and 0.95%-0.98%, respectively) in the sixth pair of leaf. However, the first, third and fourth pair of leaves showed very little variations in N, P and K content. The nutrients (N, P and K) content of leaf within 1 m of canopy height were low (1.30%-1.61% N, 0.12%-0.16% P and 0.72%-1.10% K, respectively) which gradually increased with sampling height up to 3 m (1.38%-1.71% N, 0.15%-0.20% P and 0.77%-1.18% K, respectively) and then declined marginally up to the sampling height of 5 m or above (1.51%-1.68% N, 0.14%-0.19% P and 0.73%-1.14% K, respectively).

The available N, P and K content of soil, in general, showed maximum depletion between January and May (flowering and fruit development period) which then increased in the month of July following fertilization after fruit harvest. It declined again in September and then increased between November and January. The organic carbon content of soil, however, was minimum (0.536%) in winter (January) compared with 0.627% to 0.632% in monsoon (July to September).

The top soil (0-30 cm) layer showed highest concentration of organic carbon (0.621%), available N (285.80 kg/ha), P (54.00 kg/ha) and K (136.90 kg/ha) which declined with increase in depth of sampling and it was recorded 0.583% organic carbon, 278.90 kg N/ha, 50.89 kg P/ha and 132.80 kg K/ha at 90 cm soil depth.

Considering temporal and spatial variations in soil nutrients content, it appears that a composite sample should be drawn for estimation taking soil samples randomly from atleast 10 spots (per hectare) at 0-90 cm soil depth. The sample may be drawn in the month of March for estimating soil nutrient status and fertilizer recommendation.

Considering the variations in leaf nutrients content due to the variables (direction, leaf age and canopy height) of leaf sampling, it may be suggested that a representative, composite sample of 16-20 pair of leaflets should be collected from four directions (east, west, north and south) of atleast four trees per hectare at the canopy height between 2 and 4 m from ground, taking the second pair of leaflets of the second leaf from tip of the shoot. At anthesis, the mature leaf (2nd pair) just behind the panicle should be sampled as the suitable 'index tissue' for estimating nutrition status of bearing litchi tree cv. Bombai.

The pre-flowering (December) leaf nutrient status may be estimated for prediction of flower bud differentiation and flowering intensity. If the leaf N content ranged between 1.40% and 1.60% with leaf and shoot C/N ratio ranging between 4.5-5.5 and 7.5-8.5, respectively, atleast 40% of shoots on a tree may be expected to bear panicle in the spring.

The tree and soil nutrition status should be estimated at anthesis (March) for optimum yield and quality of fruit. The available soil N content of about 300.0 kg/ha and 1.65% to 1.75% leaf N content at anthesis (March) supported a yield of 43.83 kg fruit/tree when 200 g N/tree was applied after fruit set (March) and other factors (irrigation, climate, pest and diseases) were not yield limiting. If the leaf N content at anthesis ranged between 1.65% and 1.75%, total N (400 g/tree) should be applied in two splits in July (after fruit harvest) and March (after fruit set) for optimum yield and better quality of fruit in 22 years old 'Bombai' litchi trees grown in alluvial region of West Bengal. Based on the adequate leaf nutrient ranges of N : 1.65-1.75%, P : 0.18-0.20% and K : 1.00-1.20% at anthesis, we suggest to apply N, P_2O_5 and K_2O at 400, 200 and 400 g/tree/year in two splits in July (after fruit harvest) and March (after fruit set) for a yield target of 7.80 t/ha from 22 years old 'Bombai' litchi orchard grown in alluvial region of West Bengal.

Experiment III : To study the effects of water relations and CO_2 assimilation on growth and yield of field grown trees.

In this experiment, soil moisture was supplemented from anthesis through drip irrigation system at 80, 60, 40, 20 and 0% of pan coefficient (i.e., five treatments). Basic informations were recorded on bulk density and average soil water content (θ) from 0-150 cm at field capacity. Soil water content was measured weekly with a neutron moisture probe (503 Hydroprobe, CPN, USA). Water use (E_t) in the experimental plots was calculated from the change in average soil water content from 0-150 cm each week, assuming no deep drainage below 150 cm. Leaf water potential (Ψ_L) of a single leaf per tree was measured on the shaded side of each tree at 0900 h and on the sunlit side of each tree at 1400 h with a pressure chamber. Samples were collected weekly (only from week 5 for the afternoon sampling). Stomatal conductance (gs) and net CO_2 assimilation rate (A) were measured with a photosynthesis meter (CI 310, CID, Inc.). The effect of different levels of irrigation on soil water content, shoot growth, CO_2 assimilation rate, fruit growth, sun-burning and skin-cracking of fruit, yield and quality of fruit and weekly water use, crop factor and water expense efficiency have been studied.

Application of irrigation at 80% pan coefficient (T_1) showed the highest soil water content (SWC) (20.00% to 20.93%) while lowest SWC (15.87% to 19.77%) was recorded due to dry treatment (T_5). In other words, there was the minimum fluctuations in SWC which was almost static at 93.30% of field capacity in all the weeks by irrigating at 80% pan coefficient (T_1) compared with the maximum fluctuations in SWC (90.7% of field capacity in week 1 to 72.8% of field capacity in week 9) due to dry treatment (T_5). At lower irrigation levels, the SWC fell below 90% of field capacity at week 3 (T_3, T_4 and T_5), week 5 (T_3, T_4 and T_5), week 7 (T_3, T_4 and T_5) and week 9 (T_2, T_3, T_4 and T_5).

The morning leaf water potential (LWP) was above -1.30 Mpa in all the weeks by irrigating at 60% and 80% pan coefficient (T_2 and T_1). At lower irrigation levels, the morning LWP fell below -1.30 Mpa at week 3 (T_3, T_4 and T_5), week 5 (T_4 and T_5), week 7 (T_5) and week 9 (T_3, T_4 and T_5). Irrigation at 80% pan coefficient (T_1) caused higher LWP in the afternoon (-1.44 Mpa to -1.23 Mpa) compared with afternoon LWP of -2.78 Mpa to -2.36 Mpa in dry treatment (T_5). At lower irrigation levels, the afternoon LWP fell below -2.0 Mpa at week 5 (T_4 and T_5), week 7 (T_4 and T_5) and week 9 (T_3, T_4 and T_5). The highest relative water content (RWC) of leaf (91.74 to 94.35%) was recorded due to irrigation at 80% pan coefficient (T_1) compared with the lowest (83.25% to 90.42%) in dry treatment (T_5). The decline in RWC of leaf below 90% corresponded to the LWP below -1.30 Mpa when the levels of applied irrigation decreased at different weeks.

Irrigation at 80% pan coefficient (T_1) caused the highest (39.95%) increase in total shoot growth compared with the lowest (16.98%) in dry treatment (T_5). The increase in growth due to irrigation levels, however, showed higher variations at week 9 (0.78 to 3.93%), week 5 (1.12 to 4.65%) and week 3 (2.49 to 5.93%) compared with week 1 (4.49 to 5.05%) and week 7 (1.40 to 1.81%).

The levels of irrigation (treatments) caused marked variations in stomatal conductance and net CO_2 assimilation rate in all the weeks. Irrigation at 80% pan coefficient (T_1) caused the highest stomatal conductance (26.45 to 36.84 milimolm^{-2}s^{-1}) and net CO_2 assimilation rate (7.10 to 11.90 µmolm^{-2}s^{-1}) compared with the lowest stomatal conductance (15.65 to 19.53 milimolm^{-2}s^{-1}) and net CO_2 assimilation rate (4.20 to 8.60 µmolm^{-2}s^{-1}) in dry treatment (T_5). In week 3 (post fruit set stage), the stomatal conductance fell below 25.00 milimolm^{-2}s^{-1} and the net CO_2 assimilation rate fell below 9.00 µmolm^{-2}s^{-1} due to low levels of irrigation (T_3, T_4 and T_5) treatments that corresponded with the SWC (0-150 cm depth) below 90% of field capacity and morning LWP below -1.30 Mpa. The decline in stomatal conductance and net CO_2 assimilation rate as the leaf water potential fell (below -1.30 Mpa) under dry (T_5) and partial dry (T_4, T_3) treatments was pronounced in all the weeks.

The flowering characters showed significant variations in 2004 but were non-significant in 2003. Maximum length (29.46 cm) and breadth (16.35 cm) of panicle in 2004 were recorded by irrigating at 80% pan coefficient, while it was minimum (26.90 cm and 14.82 cm, respectively) in dry treatment. The number of staminate and pistillate flowers per panicle showed no specific trend with different irrigation levels in 2003, but the number of both types of flowers per panicle in 2004 showed gradual decline with decrease in the levels of irrigation. The ratio of the number of staminate and pistillate flowers on a panicle varied between 2.53 : 1 and 2.74 : 1 in 2003 and between 2.59 : 1 and 3.01 : 1 in 2004, but showed no specific trend with irrigation levels. The number of fruits set per panicle showed gradual decrease at the lower levels of irrigation treatments in both the years. Irrigating at 80% pan coefficient caused maximum fruit set (14.65/panicle in 2003 and 20.16/panicle in 2004) compared with the lowest (13.26/panicle in 2003 and 15.01/panicle in 2004) in dry treatment.

The period between 1st and 3rd week after fruit set (WAFS) was found very sensitive to moisture stress when lower levels of irrigation (at or below 40% pan coefficient) as well as dry treatment caused more than 12.5% fruit drop at 1 WAFS and more than 18.0% fruit drop at 2 WAFS. The total drop of fruit (until maturity) was lowest (21.07%) at the highest level of irrigation (T_1:0.8 E_0) treatment, while the dry treatment caused the maximum fruit drop of 52.76 percent. The decline in net CO_2 assimilation rate as LWP fell below -1.30 Mpa under dry (T_5) and partial dry (T_4, T_3) treatments appeared to have adverse effect on retention and growth and development of fruit.

Moisture stress showed significant influence on the intensity of sun-burning and skin-cracking of fruit. The minimum intensity of sun-burning and skin-cracking were recorded by irrigating at 80% pan coefficient (3.64% and 2.20%, respectively), while it was as high as 12.95% and 9.67%, respectively in dry treatment (T_5). More than 93% of the harvested fruits were normal by application of irrigation during fruit growth and development period at 60% and 80% pan coefficient (T_2 and T_1).

Application of irrigation at 80% and 60% pan coefficient (T_1 and T_2) significantly increased the weight of fruit and aril content over the dry treatment (T_5) in both the years. The weight of fruit was minimum (21.27g to 22.23g) in both the years due to dry treatment (T_5) compared with 24.71 to 24.95g in T_1 (0.8 E_0) treatment. Moisture stress due to application of irrigation at or below 40% pan coefficient (T_3, T_4 and T_5) reduced the aril content (< 61.5%) of fruit which was significantly higher (> 63.5%) at irrigation levels of 60% and 80% pan coefficient (T_2 and T_1).

More than 2500 fruits per tree were produced on the trees that received irrigation at 60% and 80% pan coefficient however, the yield markedly reduced to less than 2250 fruits per tree at lower levels (at or below 40%) of irrigation treatments. The highest yield of 67.65 kg fruit per tree in 2003 and 71.25 kg fruit per tree in 2004 were recorded by irrigating at 80% pan coefficient (T_1) compared with 27.67 kg fruit per tree in 2003 and 32.60 kg fruit per tree in 2004 in dry treatment (T_5).

Different levels of irrigation showed significant effect on total soluble solids (TSS) content of fruit in both the years. The TSS content of fruit was recorded highest (18.87^0B in 2003 and 18.95^0B in 2004) by irrigating at 80% pan coefficient (T_1) compared with the lowest (17.83^0B in 2003 and 17.95^0B in 2004) in dry treatment (T_5). Maximum content of reducing sugar was recorded (15.12% in 2003 and 15.21% in 2004) by irrigating at 60% pan coefficient (T_2). The highest amount total sugar content of fruit was recorded (17.33%) in 2003 due to T_4 (0.2 E_0) and 17.10% in 2004 due to T_2 (0.6 E_0) treatments. Irrigating at 80% pan coefficient caused the minimum (0.42% in 2003 and 0.45% in 2004) fruit acidity which was maximum in T_5 (0.63%) in 2003 and T_3 (0.53%) treatments in 2004. An overall decrease in the TSS/acid ratio of fruit was found with the increasing intensity of moisture stress. The maximum TSS/acid ratio of 44.92 in 2003 and 42.11 in 2004 were recorded by irrigating at 80% pan coefficient.

Different irrigation levels caused marked variations in weekly water use (Et) which varied between 24.14±3.95 mm in highest level of irrigation (T_1:0.8 E_0) and 20.17±3.75 mm in dry treatment (T_5). The crop factor (Kc) of litchi cv. Bombai (average of two years) varied between 0.33 and 0.59 during the fruit growth and development period. At higher irrigation level of 80% pan coefficient (T_1), the Kc was higher (0.43 to 0.59) compared with 0.33 to 0.52 in dry treatment (T_5). The overall value of Kc between anthesis and harvest was 0.52 by irrigating at 80% pan coefficient, which declined at lower levels (T_2, T_3 and T_4) of irrigation and was found lowest (0.44) in dry treatment (T_5). The total water use (Et) of 15 years old Bombai litchi trees between anthesis and harvest was 190.50 mm with a corresponding water expense efficiency (WEE) of 452.50 kg/ha/cm. The total Et was recorded maximum (211.20 mm) by irrigating at 80% pan coefficient (T_1) and minimum (172.30 mm) in dry treatment (T_5). However, the WEE was maximum (590.18 kg/ha/cm) when irrigated at 60% pan coefficient (T_2), closely followed by 578.18 kg/ha/cm in T_1 (0.8 E_0) treatment. There was a sharp decline in WEE at lower irrigation levels (at or below 40% pan coefficient, T_3 and T_4) and it was lowest (305.94 kg/ha/cm) in dry treatment (T_5).

The benefit : cost ratio of litchi cultivation under drip irrigation (average of four irrigation levels) was 4.12 : 1. Irrigation at 80% pan coefficient (T_1) caused net return of Rs. 2,61,272.00 per hectare with a benefit : cost ratio of 5.47 : 1, while the net return was very low (Rs. 1,34,101.00) in dry treatment (T_5) resulting in a very poor benefit : cost ratio of 1.81 : 1.

From the present study it appears that adequate soil moisture during fruit growth and development period (March to May) is essential to minimize fruit drop and for better development of fruit, higher yield and improved quality of fruit. During this period, the leaf water potential in the morning should not fall below -1.30 Mpa to ensure maximum rate of stomatal conductance and net CO_2 assimilation that influence the fruit development and yield. It can be achieved when soil water content is maintained at near field capacity by irrigating at 80% pan coefficient which is equivalent to an overall crop factor of 0.52 (for irrigation scheduling) and application of irrigation to 15 years old bearing litchi trees cv. Bombai grown on sandy loam soil from anthesis to harvest (March to May) should be by drip irrigation using 2 drippers tree^{-1} discharging 8 litre water hour^{-1} dripper^{-1} for 3 hours at alternate day. The corresponding water expense efficiency and benefit : cost ratio estimated as 578.18 kg/ha/cm and 5.47 : 1, respectively.

7.0 BIBLIOGRAPHY

A.O.A.C. (1984).*Official Methods of the Analysis of the Association of Official Analytical Chemist.* Washington, D. C., 14th Edn.

Anonymous (1978).Lychee in Guangdong.*Publ.South China Agril. Univ.,* Guangdong. p. 92.

Anonymous (2001).Area and production of Fruits.*Agril. Res. Data Book,* ICAR, New Delhi. p. 92.

Anonymous (2002).Area and production of Fruits.*Agril. Res. Data Book,* ICAR, New Delhi. p.122.

Anonymous (2002-03). Area and production of Fruits and vegetable in West Bengal.*Economic Rev., Statistical Apendix.*p. 87.

Anonymous (2002).Mango, litchi research centre in Bengal soon. *Financial Daily, The Hindu,* Saturday, 5th October.

Anonymous (2003).All-India area and production of Fruits.*Agril. Res. Data Book,* ICAR, New Delhi. p.126.

Anonymous (2003-04). Area and production of Fruits and vegetable in West Bengal.*Economic Rev., Statistical Apendix.*p. 87.

Azad, K.C. (1972). The standardization of the foliar sampling technique in litchi (*Litchi chinensis* Sonn.).*Ph. D. thesis submitted to the Punjab Agric. Univ.,* Ludhiana, Punjab.

Barrs, H.D. and Weatherley, P.E. (1962).A re-examination of relative turgidity technique for estimating water deficits in leaves.*Australian J. Biol. Sci.,* **15** : 413-418.

Batten, D., Lloyd, J. and McConchie, C. (1992). Seasonal variation in stomatal response of two cultivars of lychee (*Litchi chinensis* Sonn.).*Australian J. Plant Physiol.,* 19 : 317-329.

Batten, D.J. (1984) Lychee varieties. *Department of Agric., New South Wales.Agfact.,* **H 6.2.7** : 1-15.

Bhargava, B.S. (1999). Leaf analysis for diagnosing nutrients in fruit crops.*Indian Hort.*, **43** (4) : 6-8.

Bielorai, H., Levin, I. and Assaf, R. (1973).Irrigation of fruit trees. In: *Arid Zone Irrigation* (Eds. Yaron, B. *et al.*), Chapman and Hall, London. pp.397-404.

Black, C.A. (1965). Methods of soil analysis.*American Soc. Agron.Inc.*Madison. pp. 1171-1175.

Bloom, A.J. (1992).Assimilation of mineral nutrients.In :*Plant Physiology*, 2nd edn. (eds. L. Taiz and E. Zeiger), Sinauer Association Inc., Sunderland, Masschusetts. pp. 324-343.

Bredell, G.S., Barnard, C.J. and Vincent, A.P. (1975).Application of herbicides through microjet irrigation system.*Citrus and Sub-trop. Fruit J.*,**497** : 17-18.

Chaikiattiyos, S., Menzel, C.M. and Rasmussen, T.S. (1994). Floral induction in tropical trees : effects of temperature and water supply. *J. Hort. Sci.*, **69** : 397-415.

Chandra Babu, S. (2002). National Agricultural policies : Addressing some general questions. *Hindu Survey of Indian Agric.*, M/s, Kasturi and Sons Ltd., Chennai. pp. 33-35.

Chapman, H.D. and Pratt, P.F. (1961).*Methods of Analysis for Soil, Plants and Water*. Univ. California, USA. p.6.

Chartzoulakis, K., Michelakis, N. and Tzompanakis, I. (1992).Effect of water amount and application date on yield and water utilization efficiency of 'Koroneiki' olives under drip irrigation.*Advances Hort. Sci.*, **6** (2) : 82-84.

Chaudhury, S.K. and Singh, S.N. (1993).Foliar sampling technique in litchi (*Litchi chinensis* Sonn.).*Abst. Hort. Soc. India Golden Jubilee Symp.*, Bangalore, 24-28 May, 1993.

Chen, S.T., Huang, H.Y., Lin, H.S. and Chang, L.R. (ed.) (1994).Studies on the flush growth and panicle formation of litchi.*Rept.Taichung District Agril.Improvement St.*, No.33. pp. 119-128.

Daniel, J.C. (2002). Water Balance of the Plant.In :*Plant Physiol.*, 2nd Edn. (eds. L. Taiz and E. Zeiger). Sinauer Assoc., Inc., Publ. Sunderland, Massachusetts. pp. 81-101.

Das, B., Yadav, V.B. and Mitra, S.K. (2001).Standardization of dose and time of nitrogen fertilizers in litchi (*Litchi chinensis* Sonn) cv. Bombai.*Environment and Ecology*, **19** : 516 – 518.

Davenport, T.L., Li, Y. and Zheng, Q. (1999).Towards reliable flowering of lychee (*Litchi chinensis* Sonn) in south Florida. *Proc. Florida St. Hort. Soc.*, **112** : 182-184.

Ghosh, B. and Mitra, S.K. (1990).Effect of varying level of nitrogen, phosphorus and potassium on yield and quality of litchi (*Litchi chinensis* Sonn.) cv. Bombai.*Haryana J. Hort. Sc.*, **19** (1-2) : 7-12.

Ghosh, S.P. (2001) World trade in litchi: Past, Present and future. *Acta Hort.*, **558** : 25-30.

Gogoi, S., Bhuyan, M.K. and Karmakar, R.M. (2003).Dynamics of microbial population in tea ecosystem.*Jr. Indian Soc. Soil Sci.*, **51** : 252-257.

Gomez, K.A. and Gomez, A.A. (1983).*Statistical Procedures for Agricultural Research*, 2[nd]Edn., John Willey and Sons, New York. pp. 20-29.

Gowda, Thimme and Gowda, T. (1990).A brief review of drip irrigation in Karnataka.*Proc. 11th Int. Sci. Cong. Use of Plastic in Agriculture*, February 26 to March 2, New Delhi, India, B.pp. 131-135.

Groff, G. W. (1921). *The lychee and longan.*Canton Christian College and Orrange Judd Co., New York.p. 180.

Groff, G. W. (1943). Some ecological factors involved in successful lychee. *Proc. Florida St. Hort. Soc.*, **56** : 134-155.

Guimera, J., Marfa, O., Candela, L. and Serrano, L. (1995). Nitrate leaching and strawberry production under drip irrigation management. *Agriculture, Ecosystem and Environment*, **56** : 121-135.

Hasan, M.A. and Chattopadhyay, P.K. (1993).Fruit quality of litchi under the influence of nitrogen, phosphorus and potassium nutrition.*J. Potassium Res.*, **9** : 360 – 364.

Hasan, M.A. and Chattopadhyay, P.K. (1997).Leaf nutrient status and their correlative relationship with yield in litchi cv. Bombai under the influence of N, P and K nutrition.*Indian Agric.*, **41** : 61 – 69.

Hassan, M. A. and Chattopadhyay, P.K. (1990).Effect of different soil moisture regimes on growth and yield of litchi cv. Bombai.*Indian J. Hort.*, **47** : 397-400.

Hassan, M.A. and Chattopadhyay, P.K. (1991). Flowering behavior of litchi as affected by different soil moisture regimes. *South Indian Hort.*, 39 (2) : 67-70.

Hassan, M.A. and Chattopadhyay, P.K. (1992). Note on effect of different soil moisture regimes on water requirement, crop coefficient and water use efficiency of litchi. *Indian J. Hort.*, **49** : 155-157.

Hayes, W.B. (1957). *Fruit Growing in India*, 2nd Edn. Kitabistan, Allahabad. pp.

Hedge, D.M. and Srinivas, K. (1990). Drip irrigation is recommended for water-consuming banana. *Indian Hort.*, **35** (2) : 4-6.

Heinicke, D.R. (1976). The sun shine.*American Fruit Grower*, **96** : 33-52.

Hieke, S., Menzel, C.M., Doogan, V.J. and Ludder, P. (2002a).The relationship between fruit and leaf growth in lychee (*Litchi chinensis* Sonn.)*J. Hort. Sci. Biotech.*, **77** : 320-325.

Hieke, S., Menzel, C.M., Doogan, V.J. and Ludder, P. (2002b). The relationship between yield and assimilate supply in lychee (*Litchi chinensis* Sonn.). *J. Hort. Sci. Biotech.*, **77** : 326-332.

Hillel, D. (1987). The efficient use of water in irrigation.*The World Bank Tech. Paper***No.64**, World Bank, Washington D.C. p. 101.

Huang, W.T., Huang, Y.M., Hsiang, W.M., Chang, M.H., Lin, M.L., Wang, C.H. and Wu, W.L. (1998). Nutrition studies on lychee (*Litchi chinensis* Sonn) orchards in central Taiwan. *J. Agric. Res. China*, **47** : 388 – 407.

Hundal, H.S. and Arora, C.L. (1996). Preliminary micronutrients foliar diagnostic norms for litchi (*Litchi chinensis* Sonn.) using DRIS. *Jr. Indian Soc. Soil Sci.*, **44:** 294-298.

Jackson, M.L. (1967). *Soil Chemical Analysis.*Prentice Hall, India.

Kadman, A. and Slor, E. (1982).Litchi growing in Israel.*Alon Hanotea*, **36** : 673-688.

Kanwar, J.S. (2002). *Litchi – maturity, harvesting, storage, marketing and yield.* Kalyani Publ., New Delhi. pp.188-215.

Kanwar, J.S. and Nijjar, G.S. (1975).Litchi Cultivation in Punjab.*Punjab Hort. J.*, **15**: 8-13.

Kanwar, J.S., Nijjar, G.S. and Rajput, M.S. (1972). Fruit growth studies in litchi (*Litchi chinenis* Sonn.) at Gurdaspur. *Haryana J. Hort. Sci.*, **12**: 146-151.

Koen, T.J. and Smart, G. (1982).Effect of optimal manure on the production and fruit quality of litchi trees.*Inf. Bull. Citrus and Sub-tropical Fruit Res. Inst.*, **171** : 1 – 2.

Koen, T.J. and Smart, G. (1983a). Fertilization of litchis.*Inf. Bull. Citrus and Sub-tropical Fruit Res. Inst.*, p. 1.

Koen, T.J. and Smart, G. (1983b). Leaf Analysis in litchi.*Inf. Bull. Citrus and Sub-tropical Fruit Res. Inst.*, p. 2.

Koen, T.J., Langengger, W. and Smart, G. (1981).Determination of the fertilizer requirements of litchi trees.*Inf. Bull. Citrus and Sub-tropical Fruit Res. Inst.*, **103** : 9 – 12.

Kotur, S. C. and Singh, H. P. (1992a).Varietal differences in leaf nutrient composition in litchi (*Litchi chinensis* Sonn.)*Abst. Nat. Semi. Recent Developments in litchi cultivation*, Rajendra Agric. Univ., Pusa (Samastipur), Bihar. pp. 25-26.

Kotur, S. C. and Singh, H. P. (1992b).Studies on leaf sampling of litchi (*Litchi chinensis* Sonn.)*Abst. Nat. Semi. Recent Developments in litchi cultivation,* Rajendra Agric. Univ., Pusa (Samastipur), Bihar.p. 23.

Kumar, A.P.A. and Bojappa, K.M. (1994).Studies on the effect of drip irrigation on yield and quality of fruits in sweet oranges (*Citrus sinensis* (L) Osbeck) and economy in water use.*Mysore J. Agril. Sci.*, **28** : 338-344.

Lagatu, H. and Maume, L. (1926).Diagnostic di'. Alimentation d'un vegetal par I' evolution chemique d'une feuille convenablement choisie, *C. R. Acad. Sci. Paris*, **182** : 635-655.

Langenegger, W. (1974). Nutrient requirement of litchi: Leaf and soil analysis. *Inf. Bull.Citrus and Sub-tropical fruit Res. Inst.*, **20** : 9 – 10.

Langenegger, W. (1975).Litchi nutrient trial.*Inf. Bull.Citrus and Sub-tropical fruit Res. Inst.*, **37** : 5 – 6.

Lawlor, D.W. (2001). Photosynthesis by leaves - water stress. In: *Photosynthesis*, 3[rd]Edn., Viva Books Pvt. Ltd., New Delhi. pp. 309-351.

Li, Jian Guo, Chen, H. Li, J.G. and Chen, H. (1998). Study on the effect of sprouting period of last autumn shoot on the flower quality and fruit set of litchi cv. Nuomizi. *South China Fruits*, **27** (4) : 24-25.

Li, Jian Guo, Gao, Fei Fei, Huang, Hui Bai, Tan, Yao Wen, Luo, Jing Tang, Li, J.G., Gao, F.F., Huang, H.B., Tan, Y.W. and Luo, J.T. (1999). Preliminary studies on the relationship between calcium and fruit cracking in litchi fruit.*J.South China Agril. Univ.*, **20** (3) : 45-49.

Li, Jian Guo, Huang, Hui Bai, Li, J.G., and Huang, H.B. (1995). Physioco-chemical properties and peel morphology in relation to fruit cracking susceptibility in litchi fruit. *J. South China Agril. Univ.*, **16** (1) : 84-89.

Maity, S.C. and Mitra, S.K. (2001).Litchi. In: *Fruits: Tropical and subtropical*, **Vol. I** (eds. T.K. Bose, S.K. Mitra and D. Sanyal) Naya Udyog, Calcutta, India. pp. 556-608.

Marchal, J. (1984).Miscellaneous Tropical Fruits. In : *Plant Analysis – as a guide to the Nutrient requirement of Temperate and Tropical Crops* (eds. P. Martin-Prevel, J. Gagnard and P. Gautier), SBA Publication, Calcutta, India. pp. 430 – 433.

Martin-Prevel P., Gagnard, J. and Gautier, P. (1984).*Plant Analysis – as a guide to the Nutrient requirement of Temperate and Tropical Crops.*SBA Publication, Calcutta.p. 77.

Menzel, C.M. (1984). Management of bearing lychee tree in sub-tropical Queensland.*Newst.Sunshine Coast Subtrop. Fruits Association*, **9** : 20 – 31.

Menzel, C.M. and Simpson, D.R. (1986a). Lychee production around the world. (eds. C.M. Menzel and G.N. Greer). *Proc. 1ˢᵗ Nat .Lychee Semi.*, Sunshine Coast Subtrop. Fruits Assoc., Nambour, Queensland, Austrilia. pp. 55-70.

Menzel, C.M. and Simpson, D.R. (1986b). The role of potassium in lychee flowering. *Rept. Maroochy Hort. Res. Station,*4 : 76.

Menzel, C.M. and Simpson, D.R. (1986c). Plant water relation in lychee : Diurnal variation in leaf conductance and leaf potential. *Agril. Forest Meteorology*, **37** : 267-277

Menzel, C.M. and Simpson, D.R. (1987).Lychee nutrition – a review.*Scientia Hort.*, **31** : 195 – 224.

Menzel, C.M. and Simpson, D.R. (1990). Nutritional studies on lychee trees in subtropical Austrilia. *Acta Hort.*, **275** : 581 – 585.

Menzel, C.M. and Simpson, D.R. (1992).Growth, flowering and yield of litchi cultivars.*Scientia Hort.*, **49** : 243-254.

Menzel, C.M., Atiken, R. L., Dowling, A. W. and Simpson, D.R. (1990). Root distribution of litchi trees growing in acid soils of subtropical Queensland. *Australian J. Exp. Agric.*, **30** : 699-705.

Menzel, C.M., Carseldine, M.L. and Simpson, D.R. (1987).The effect of leaf age on nutrient composition of non-fruiting litchi (*Litchi chinensis* Sonn.).*J. Hort. Sc.*, **62** : 273-279.

Menzel, C.M., Carseldine, M.L. and Simpson, D.R. (1988). The effect of fruiting status on composition of litchi (*Litchi chinensis* Sonn) during the flowering and fruiting season. *J. Hort. Sci.*, **63** :   547 – 556.

Menzel, C.M., Haydon, G.F and Simpson, D.R. (1992a). Mineral nutrient reserves in bearing litchi trees (*Litchi chinensis* Sonn.).*J. Hort. Sci.*, **67** : 149-160.

Menzel, C.M., Carseldine, M.L., Haydon, G.F. and Simpson, D.R. (1992b) A review of existing and proposed new leaf nutrient standards for lychee. *Scientia Hort.*, **49** : 33-53.

Menzel, C.M., Haydon, G.F., Doogan,V.J. and Simpson, D.R. (1992c) Observation on the leaf nutrient status in Australia. *J. South African Soc. Hort. Sci.*, **2** (2) : 86-88.

Menzel, C.M., Haydon, G.F., Doogan, V.J. and Simpson, D.R. (1994).Time of nitrogen application and yield of Bengal lychee on a sandy loam soil in subtropical Queensland.*Australian J. Expt. Agric.*, **34** : 803 – 811.

Menzel, C.M., Oosthuizen, J.H., Roe, D.J. and Doogan, V.J. (1995a). Water deficits at anthesis reduce CO_2 assimilation and yield of lychee (*Litchi chinensis* Sonn.) trees. *Tree Physiol.*, **15** : 611-617.

Menzel, C.M., Rasmussen, T. S. and Simpson, D.R. (1995b). Carbohydrate reserves in lychee trees (*Litchi chinensis* Sonn.). *J. Hort. Sci.*, **70** : 245-255.

Menzel, C.M., Simpson, D.R., Haydon, G.F. and Doogan, V.J. (1995c). Phosphorus and potassium fertilization of lychee.*Jr. South African Soc. Hort. Sci.*, **5** : 97 – 99.

Mitra, S.K. (2004a). Litchi cultivation in China.*Paper presented at the Nat. Semi. "Advances in production and post harvest technology of litchi for export"*, June 24-26, BCKV, West Bengal.

Mitra, S.K. (2004b). Sustainable litchi production in West Bengal, India.*Acta Hort.,***632** : 64-68.

Morgan, J.M. (1984). Osmoregulation and water stress in higher plants. *Ann. Rev. Plant Physiol.*, **35** : 299-348.

Mustard, M.J. and Lynch, S.J. (1959).Notes on litchi panicle development.*Proc. Florida St. Hort. Soc.*, **72** : 324-327.

Neumann, H.H., Thurtell, G.W. and Slevenson, K.R. (1974).Leaf water content and potential in corn, sorghum, soybean and sunflower.*Canadian J. Plant Sci.*, **54** : 175-184.

Nijjar, G.S. (1981). *Litchi cultivation.*Punjab Agric. Univ. Ludhiana.p. 36.

Njoku, E. (1957). The effect of mineral nutrition and temperature on leaf shape in *Ipomoea carcrulca.New Phytol.*,**56** : 154-171.

Oosthuizen, J.H., Roe, D.J. and Menzel, C.M. (1994). Response of litchi to drought : water stress, photosynthesis, growth and yield of trees growing in a sandy loam soil. *Year Book South African Litchi Growers' Assoc.*, **6**: 24-28.

Patel, V. B. (1999). Using water sensibly.*Hindu Survey of Indian Agric.* p. 165.

Paxton, B. and Chapman, K.R. (1980). Some litchi yields of interest at Maroochy Horticultural Research Station. *Bicnnial Rep. Maroochy Hort. Res. St.*, **2** : 344-347.

Plaut, Z. and Moreshet, S. (1973). Transport of water in plant atmosphere system. In: *Arid Zone Irrigation* (eds. Yaron, B. *et al.*) Chapman Hall, London. pp. 123-141.

Rao, D.P., Mukherjee, S.K. and Ray, R.N. (1985). Nutritional studies of litchi orchards in different districts of west Bengal. *Indian J. Hort.*, **42** : 1-7.

Ray, P.K. (1991a). Fertilizer use for proper nutrition of litchi orchards.*Adhunik Kisan*, **21** (7) : 9-12.

Ray, P.K. (1991b). Irrigation of litchi orchards.*Adhunik Kisan*, **21** (7) : 12-14.

Ray, P.K. (2001a). Origin, distribution and economic importance.In :*Litchi – Botany, Production and utilization* (ed. K.S. Chauhan), Kalyani Publ. pp.2-9.

Ray, P.K. (2001b). Soil and plant nutrition.In :*Litchi – Botany, Production and utilization* (ed. K.S. Chauhan), Kalyani Publ. pp.76-97.

Ray, P.K., Mishra, K.A. and Sharma, S.B. (1985).Yield and bearing constancy of some commercial litchi cultivars.*Indian J. Hort.*, **42** : 218 – 222.

Rivera, L.J., Ordorica, F.C., Wesche, E.P. (1999). Changes in anthocyanin concentration in lychee (*Litchi chinensis* Sonn.) pericarp during maturity.*Food Chemistry*, **65** : 195-200.

Robson, A.D. (1981). Principal and approaches used in plant analysis. *Proc. Nat. Workshop Plant Analysis*, pp. 1-7.

Roe, D.J. and Oosthuizen, J.H. (1994).Drought strategies for litchi orchards.*Year Book South African Litchi Growers' Assoc.*, **6** : 29-31.

Roe, D.J., Menzel, C.M., Oosthuizen, J.H. and Doogan, V.J. (1997).Effect of current CO_2 assimilation and stored reserves on lychee fruit growth.*J. Hort. Sci.*, **72** : 397-405.

Roe, D.J., Oosthuizen, J.H. and Menzel, C.M. (1995).Rate of soil drying and previous water deficits influence the relationship between

CO_2 assimilation and tree water status in potted lychee (*Litchi chinensis* Sonn.).*J. Hort. Sci.*, **70** : 15-24.

Salisbury, F.B. and Ross, C.W. (2003a). Stress physiology - mechanisms of plant to water and related stress. In :*Plant Physiol.* CBS Publ. and Distributors, New Delhi. pp. 468-489.

Salisbury, F.B. and Ross, C.W. (2003b). The photosynthesis and transpiration compromise.In :*Plant Physiol.* CBS Publ. and Distributors, New Delhi. pp. 54-74.

Sanyal, D. and Mitra, S.K. (1990). Standardization of leaf sampling technique for mineral composition of litchi cv. Bombai. *Indian J. Hort.*, **47** : 371 – 375.

Sanyal, D., Hasan A., Ghosh. B., Mitra, S.K. (1990). Studies on sun burning and skin cracking in some varieties of litchi. *Indian Agric.*,**34** : 19-23.

Sayal, N.A., Sayal, O.U. and Jatoi, S.A. (1999). Morphology of litchi fruit as effected by exposure to sunlight and fruit orientation. *Pakistan J. Biol. Sci.*, **2** : 900-902.

Scholander, P.F., Hammel, H. T., Bradstreet, E. D. and Hemingsem E. A. (1965). Sap pressure in vascular plants. *Science*, **48** : 339-345.

Sharma, S.B. and Ray, P.K. (1987).Fruit cracking in litchi-a review.*Haryana J. Hort. Sc.*, **16** : 11-15.

Shirgure, P.S., Srivastava, A.K. and Singh, S. (2001). Effect of microjet irrigation system on growth, yield and fruit quality of Nagpur mandarin (*Citrus reticulata* Blanco).*South Indian Hort.*, **49** : 357-358.

Singh, H.P., Samuel, J.C. and Kumar, A. (2000).Micro-irrigation in horticultural crops.*Indian Hort.*, **45** (1) : 37-43.

Singh, S.N. and Pathak, R.A. (1983).Effect of irrigation intensity, its frequency and manures on the growth of litchi cv. Calcuttia.*Punjab Hort. J.*, **23** : 197-202.

Singhal, V. (1999).Indian Agriculture.*Indian Economic Data Res. Centre,* B-713, Panchavati, Vikaspuri, New Delhi.pp.159-160.

Srinivas, K. (1996). Plant water relations, yield andwater use of papaya (*Carica papaya* L.) at different evaporation-replenishment rates under drip irrigation. *Trop. Agric.*, **73** : 264-269.

Stern, R.A., Adato, I. and Gazit, S. (1990). Autumn water stress as means of increasing flowering and improving fertility in young litchi trees.*Alon Hanotea*, **44** : 391-394.

Stern, R.A., Adato, I., Goren, M., Eisentein, D. and Gazit, S. (1993). Effect of autumnal water stress on litchi flowering and yield in Israel. *Scientia Hort.*, **54** : 295-302.

Stern, R. A., Meron, M., Noar, A., Wallach, R., Bravdo, B. and Gazit, S. (1998). Effect of fall irrigation in 'Mauritius' and 'Floridian' lychee on soil and plant water status, flowering intensity and yield.*Jr. American Soc. Hort. Sci.,* **123** : 150-155.

Subbiah, B.V. and Asija, G.L. (1956).A rapid procedure for the estimation of available nitrogen in soils.*Curr. Sci.*, **25** : 259.

Syvertsen, J.P. (1985). Integration of water stress in fruit trees.*HortSci.*,**20** : 1039-1043.

Turner, N.C. and Jones, M.M. (1980). Turgor maintenance by osmotic adjustment : A review and evaluation. In: *Adaptation of plants to water and high temperature stress* (eds. N.C. Turner and P.J. Kramer), Wiley-Interscience, New York. pp. 87-103.

Wagner, M., Figueroa, M. and Laborem, G. (1984).Effect of three irrigation frequencies on the performance of mangoes (*Mangifera indica* L.) cv. Kent.*Agronomia Tropical*, **34** : 155-165.

Winks, C.W., Batten, D.J. and Burt, J.R. (1983). Australian Sub-tropical Horticulture Mission to the people's Republic of China. Commonwealth Dep. Primary Ind., Canberra.

Yee, W. (1972).*The Lychee in Hawaii*. Univ. Hawaii, Cooperative Extn. Service Circular.p. 386.

Yoshida, S., Forno, D.A., Cook, J.H. and Gomej, K.A. (1972). *Laboratory Manual for Physiological Studies of Rice*. IRRI, p. 13.

Young, T.W. and Koo, R.C.J. (1964). Influence of nitrogen source and rate of fertilization on the performance of Brewster lychees. *Proc. Florida St. Hort. Soc.*, **88** : 406-410.

Zhuang, W., Jenah, Y., Lin, Z., Chen, R. and Chen, J. (1988). A resume of research conducted on the subject of how to promote fruit bearing in mature lychee tree. *J. Fujian Agric. College*, **12** : 297 – 305.

Lightning Source UK Ltd.
Milton Keynes UK
UKHW031026190922
409092UK00001B/53